QA
433
.J3613
2001

D1257182

Undergraduate Texts in Mathematics

Editors

S. Axler
F.W. Gehring
K.A. Ribet

Springer
New York
Berlin
Heidelberg
Barcelona
Hong Kong
London
Milan
Paris
Singapore
Tokyo

Undergraduate Texts in Mathematics

Anglin: Mathematics: A Concise History and Philosophy.
Readings in Mathematics.

Anglin/Lambek: The Heritage of Thales.
Readings in Mathematics.

Apostol: Introduction to Analytic Number Theory. Second edition.

Armstrong: Basic Topology.

Armstrong: Groups and Symmetry.

Axler: Linear Algebra Done Right. Second edition.

Beardon: Limits: A New Approach to Real Analysis.

Bak/Newman: Complex Analysis. Second edition.

Banchoff/Wermer: Linear Algebra Through Geometry. Second edition.

Berberian: A First Course in Real Analysis.

Bix: Conics and Cubics: A Concrete Introduction to Algebraic Curves.

Brémaud: An Introduction to Probabilistic Modeling.

Bressoud: Factorization and Primality Testing.

Bressoud: Second Year Calculus.
Readings in Mathematics.

Brickman: Mathematical Introduction to Linear Programming and Game Theory.

Browder: Mathematical Analysis: An Introduction.

Buskes/van Rooij: Topological Spaces: From Distance to Neighborhood.

Callahan: The Geometry of Spacetime: An Introduction to Special and General Relavitity.

Carter/van Brunt: The Lebesgue–Stieltjes Integral: A Practical Introduction

Cederberg: A Course in Modern Geometries. Second edition.

Childs: A Concrete Introduction to Higher Algebra. Second edition.

Chung: Elementary Probability Theory with Stochastic Processes. Third edition.

Cox/Little/O'Shea: Ideals, Varieties, and Algorithms. Second edition.

Croom: Basic Concepts of Algebraic Topology.

Curtis: Linear Algebra: An Introductory Approach. Fourth edition.

Devlin: The Joy of Sets: Fundamentals of Contemporary Set Theory. Second edition.

Dixmier: General Topology.

Driver: Why Math?

Ebbinghaus/Flum/Thomas: Mathematical Logic. Second edition.

Edgar: Measure, Topology, and Fractal Geometry.

Elaydi: An Introduction to Difference Equations. Second edition.

Exner: An Accompaniment to Higher Mathematics.

Exner: Inside Calculus.

Fine/Rosenberger: The Fundamental Theory of Algebra.

Fischer: Intermediate Real Analysis.

Flanigan/Kazdan: Calculus Two: Linear and Nonlinear Functions. Second edition.

Fleming: Functions of Several Variables. Second edition.

Foulds: Combinatorial Optimization for Undergraduates.

Foulds: Optimization Techniques: An Introduction.

Franklin: Methods of Mathematical Economics.

Frazier: An Introduction to Wavelets Through Linear Algebra.

Gamelin: Complex Analysis.

Gordon: Discrete Probability.

Hairer/Wanner: Analysis by Its History.
Readings in Mathematics.

Halmos: Finite-Dimensional Vector Spaces. Second edition.

Halmos: Naive Set Theory.

Hämmerlin/Hoffmann: Numerical Mathematics.
Readings in Mathematics.

Harris/Hirst/Mossinghoff: Combinatorics and Graph Theory.

(continued after index)

BELL LIBRARY - TAMU-CC

Klaus Jänich

Vector Analysis

Translated by Leslie Kay

With 108 Illustrations

Springer

Klaus Jänich
NWF-I Mathematik
Universität Regensburg
D-93040 Regensburg
Germany

Leslie Kay (*Translator*)
Department of Mathematics
Virginia Tech
Blacksburg, VA 24061-0123
USA

Editorial Board

S. Axler
Mathematics Department
San Francisco State University
San Francisco, CA 94132
USA

F.W. Gehring
Mathematics Department
East Hall
University of Michigan
Ann Arbor, MI 48109
USA

K.A. Ribet
Mathematics Department
University of California at Berkeley
Berkeley, CA 94720-3840
USA

Mathematics Subject Classification (2000): 57-01, 57Rxx

Library of Congress Cataloging-in-Publication Data
Jänich, Klaus.
[Vektoranalysis. English]
Vector analysis / Klaus Jänich.
p. cm. — (Undergraduate texts in mathematics)
Includes bibliographical references and index.
ISBN 0-387-98649-9 (alk. paper)
1. Vector analysis. I. Title. II. Series.
QA433.J3613 2000
515'.63—dc21 99-16555

Printed on acid-free paper.

This book is a translation of the second German edition of *Vektoranalysis*, by Klaus Jänich, Springer-Verlag, Heidelberg, 1993.

© 2001 Springer-Verlag New York, Inc.
All rights reserved. This work may not be translated or copied in whole or in part without the written permission of the publisher (Springer-Verlag New York, Inc., 175 Fifth Avenue, New York, NY 10010, USA), except for brief excerpts in connection with reviews or scholarly analysis. Use in connection with any form of information storage and retrieval, electronic adaptation, computer software, or by similar or dissimilar methodology now known or hereafter developed is forbidden.
The use of general descriptive names, trade names, trademarks, etc., in this publication, even if the former are not especially identified, is not to be taken as a sign that such names, as understood by the Trade Marks and Merchandise Marks Act, may accordingly be used freely by anyone.

Production managed by Jenny Wolkowicki; manufacturing supervised by Erica Bresler.
Typeset by Integre Technical Publishing Co., Inc., Albuquerque, NM.
Printed and bound by Hamilton Printing Co., Rensselaer, NY.
Printed in the United States of America.

9 8 7 6 5 4 3 2 1

ISBN 0-387-98649-9 SPIN 10696691

Springer-Verlag New York Berlin Heidelberg
A member of BertelsmannSpringer Science+Business Media GmbH

Preface to the English Edition

Addressing the English-speaking readers of this book, I should state who I imagine those readers are. The preface to the first German edition was written for students in a different academic system, and the description I gave there doesn't apply directly. Should we, in this global age, have more compatibility in academic education? There is a debate going on now in Germany about whether we should introduce the bachelor's degree, or "Bakkalaureus" as some would call it, so that our system can be more easily compared with those abroad. Difficult questions! But it has been observed that whatever the academic system, students of the same *age* have about the same level of knowledge and sophistication. Therefore I can simply say that this is a book for twenty-year-old students.

This book is about manifolds, differential forms, the Cartan derivative, de Rham cohomology, and the general version of Stoke's theorem. This theory contains classical vector analysis, with its gradient, curl, and divergence operators and the integral theorems of Gauss and Stokes, as a special case. But since the student may not immediately recognize this fact, some care is given to the translation between these two mathematical languages.

Speaking of translation, I would like to acknowledge the excellent work of Leslie Kay in translating the German text into English. We have exchanged detailed e-mail messages throughout the translation process, discussing mathematics and subtleties of language. While I was using the opportunity of this English edition to eliminate all the typos and mistakes I knew of in the present German edition, Dr. Kay initiated many additional improvements. I wish to thank her for all the care she has devoted to the book.

Langquaid, Germany Klaus Jänich
October 2000

Preface to the First German Edition

An elegant author says in two lines what takes another a full page. But if a reader has to mull over those two lines for an hour, while he could have read and understood the page in five minutes, then—for this particular reader—it was probably not the right *kind* of elegance. It all depends on who the readers are.

Here I am writing for university students in their second year, who know nothing yet about manifolds and such things, but can feel quite satisfied if they have a good overall understanding of the differential and integral calculus of one and several variables. I ask other possible readers to be patient from time to time. Of course, I too would like to combine both kinds of elegance, but when that doesn't work I don't hesitate to throw line-saving elegance overboard and stick to time-saving elegance. At least that's my intention!

Introductory textbooks are usually meant "to be used in conjunction with lectures," but even this purpose is better served by a book that can be understood on its own. I have made an effort to organize the book so that you can work through it on a desert island, assuming you take your lecture notes from your first two semesters along and—in case those lectures didn't include topology—a few notes on basic topological concepts.

Since discussion partners are sometimes hard to find on desert islands, I have included *tests*, which I would like to comment on now. Some people disapprove of multiple-choice tests on principle because they think putting check marks in boxes is primitive and unworthy of a mathematician. It's hard to argue with that! Actually, some of my test questions are so utterly and obviously simple that they'll give you—a healthy little scare when you find you can't answer them after all. But many of them are hard, and resisting the specious arguments of the wrong answers takes some firmness. The tests should be taken seriously as a training partner for the reader who is alone with the book. By the way, there is at least one right answer in each set of three, but there may be several.

Now I won't describe the book any further—it's in front of you, after all—but will turn instead to the pleasant duty of looking back when the work is done and gratefully acknowledging the many kinds of help I received.

Martina Hertl turned the manuscript into TEX, and Michael Prechtel was always there with his advice and support as a TEX wizard. I received useful macros from Martin Lercher as well as from the publisher, and I was one of the first to use diagram.tex, developed by Bernhard Rauscher, for the diagrams. My students Robert Bieber, Margarita Kraus, Martin Lercher, and Robert Mandl expertly proofread the next to the last version of the book. I am very grateful for all their help.

Regensburg, Germany Klaus Jänich
June 1992

Contents

CHAPTER 1 Differentiable Manifolds

1.1 The Concept of a Manifold

The only background we need is a little topology—Chapter I of [J:*Top*] is enough, at least for now—and differential calculus of several variables.

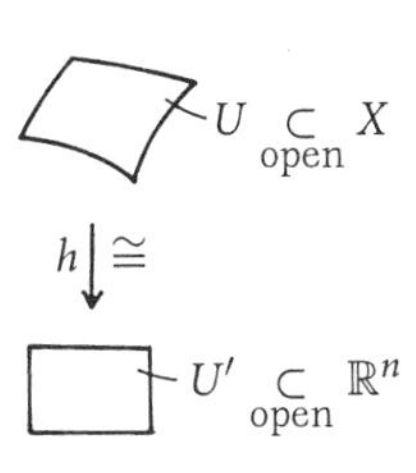

Figure 1.1. A chart

Definition. Let X be a topological space. An ***n-dimensional chart*** on X is a homeomorphism $h : U \xrightarrow{\cong} U'$ from an open subset $U \subset X$, the ***chart domain***, onto an open subset $U' \subset \mathbb{R}^n$.

If every point in X belongs to some chart domain of X, the space X is called ***locally Euclidean:*** a nice property, which of course not every topological space has.

It is often useful to include the name of the chart domain in the notation for the chart and speak of the chart (U, h), and we do so now.

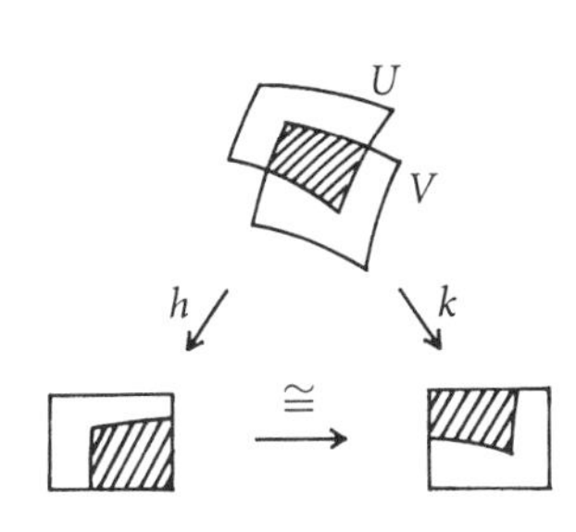

Figure 1.2. Transition map

Definition. If (U, h) and (V, k) are two n-dimensional charts on X, then the homeomorphism $k \circ (h^{-1}|h(U \cap V))$ from $h(U \cap V)$ to $k(U \cap V)$ is called the ***change-of-charts map***, or ***transition map***, from h to k. If it is not only a homeomorphism but a diffeomorphism, we say that the two charts are ***differentiably related***.

By *differentiable*, in the sense of analysis in $\mathbb{R}^n$, we always mean of class C^∞: having continuous partial derivatives of all orders. In particular, a homeomorphism f between open sets in $\mathbb{R}^n$ is a diffeomorphism if and only if both f and f^{-1} are C^∞ functions.

Definition. A set of n-dimensional charts on X whose chart domains cover all of X is an ***n*-dimensional atlas** on X. The atlas is ***differentiable*** if all its charts are differentiably related, and two differentiable atlases $\mathfrak{A}$ and $\mathfrak{B}$ are ***equivalent*** if $\mathfrak{A} \cup \mathfrak{B}$ is also differentiable.

This brings us almost to the concept of a differentiable manifold. Now we have to choose between two commonly used definitions. A *differentiable structure* on X is regarded sometimes as an equivalence class of differentiable atlases and sometimes as a maximal differentiable atlas. We first clarify in what sense the two mean the same thing.

For an n-dimensional differentiable atlas $\mathfrak{A}$, let $[\mathfrak{A}]$ denote its equivalence class and $\mathcal{D}(\mathfrak{A})$ the set of all the charts (U, h) on X that are differentiably related to all the charts in $\mathfrak{A}$. There is a differentiable transition map between any two elements of $\mathcal{D}(\mathfrak{A})$, as can be checked using $\mathfrak{A}$-charts. (This is the same argument we would have to make in verifying that "equivalence" really does define an equivalence relation on the set of atlases.) The set of charts $\mathcal{D}(\mathfrak{A})$ is thus an n-dimensional differentiable atlas and in fact obviously a *maximal* one: every chart we could have added without destroying differentiability is already there anyway. But $\mathcal{D}(\mathfrak{A})$, clearly the only maximal n-dimensional differentiable atlas that contains $\mathfrak{A}$, carries exactly the same information as the equivalence class $[\mathfrak{A}]$, because $[\mathfrak{A}]$ is just the set of all the subatlases of $\mathcal{D}(\mathfrak{A})$ and $\mathcal{D}(\mathfrak{A})$ is the union of all the atlases in $[\mathfrak{A}]$. Which to take as the structure defined by $\mathfrak{A}$ is therefore a question of taste, and I for one prefer the maximal atlas, since that is at least still an atlas:

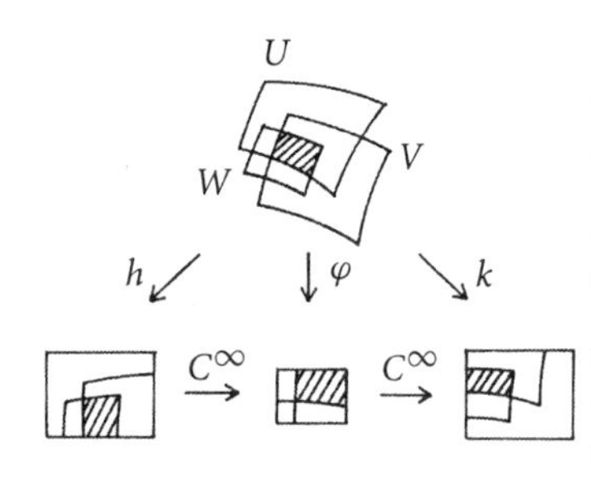

Figure 1.3. Proof of differentiability for the transition map from h to k by means of an auxiliary chart (W, φ) in $\mathfrak{A}$

Definition. An ***n-dimensional differentiable structure*** on a topological space X is a maximal n-dimensional differentiable atlas.

You probably expect a differentiable manifold to be defined as a topological space equipped with a differentiable structure, and this is essentially what is done, but two additional *topological* demands are made on the space. The first is that M must be Hausdorff, and the second that M must be *second countable*; that is, there must exist a countable basis for the topology (see, for example, [J:*Top*], p. 12 and p. 85).

Definition. An ***n-dimensional differentiable manifold*** is a pair $(M, \mathcal{D})$ consisting of a Hausdorff space M that satisfies the second axiom of countability and an n-dimensional differentiable structure $\mathcal{D}$ on M.

We usually suppress the structure in the notation and write M for a manifold, just as we write G rather than $(G, \cdot)$ for a group.

We fix a convention for the *empty* topological space with the empty structure by letting it be a manifold of any dimension, even negative. But any nonempty manifold has a well-defined dimension $n = \dim M \geq 0$.

Since we haven't defined non-differentiable manifolds and will have no need to consider them, we don't have to tack on the adjective "differentiable" every time. We also agree that, without an explicit statement to the contrary, a chart (U, h) *on the manifold* M always means a chart in the differentiable structure.

1.2 Differentiable Maps

Now we turn our attention to *maps*. Let M be a manifold, X some space, and $f : M \to X$ a map whose behavior we want to study near a point $p \in M$. Then we can choose a ***chart around p***, i.e. a chart (U, h) for M with $p \in U$, and use it to "pull the map f down." That is, we consider $f \circ h^{-1} : U' \to X$. Whatever properties and data $f \circ h^{-1}$ has locally at $h(p)$,

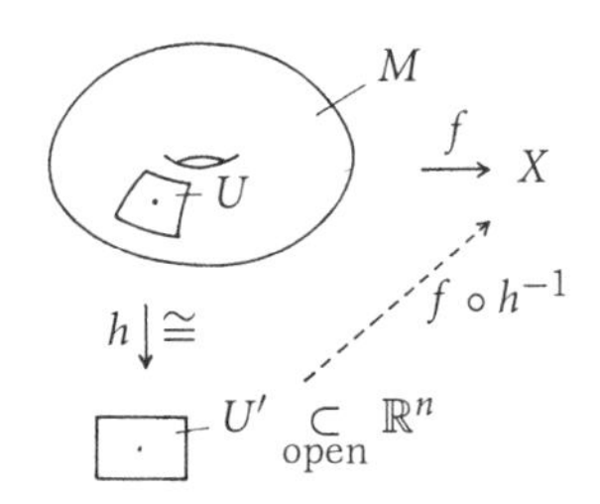

Figure 1.4. The downstairs map $f \circ h^{-1}$

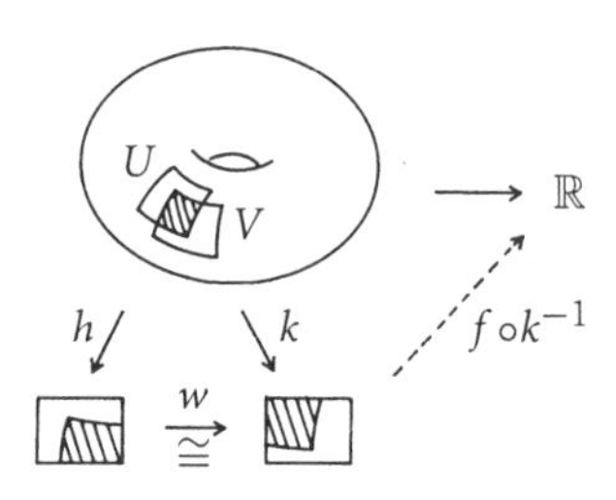

Figure 1.5. $f \circ h^{-1}$ and $(f \circ k^{-1}) \circ w$ agree on $h(U \cap V)$

we say that f has them at p ***relative to the chart*** (U, h). If such a property or datum of the "downstairs map" is actually *independent* of the choice of chart around p, so that f has the property relative to *every* chart around p, we just say that f has this property at p. For example:

Definition. A function $f : M \to \mathbb{R}$ is ***differentiable*** (i.e. C^∞) at $p \in M$ if for some (hence every!) chart (U, h) around p, the downstairs function $f \circ h^{-1}$ is differentiable in a neighborhood of $h(p)$.

The local C^∞ property at p is independent of the choice of chart because the downstairs functions relative to the charts (U, h) and (V, k) differ only by precomposition with a diffeomorphism, namely the transition map w. We proceed similarly when the target space is a manifold, but then we always start by assuming that f is continuous, since this makes possible a suitable choice of chart:

Note. *If $f : M \to N$ is a continuous map between manifolds, $p \in M$, and (V, k) is a chart around $f(p)$, then there is always a chart (U, h) around p with $f(U) \subset V$.*

In this case too, we say that f has a local property at p relative to the charts (V, k) and (U, h) if the map $k \circ f \circ h^{-1} : U' \to V'$ "pulled down" by the charts (another downstairs map) has the property at $h(p)$.

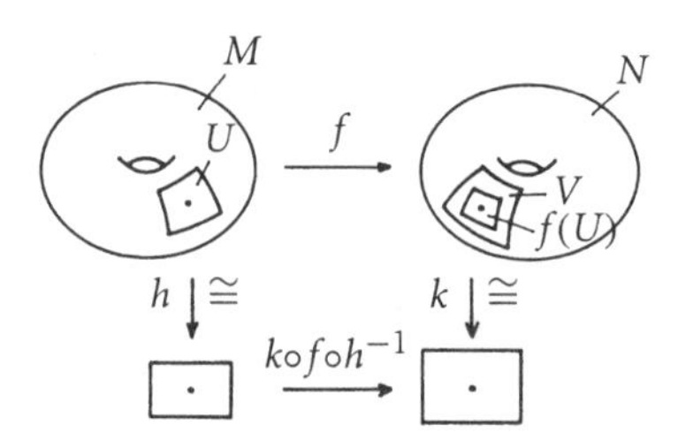

Figure 1.6. Using charts to pull down a continuous map between manifolds

Since this is a map between open sets in Euclidean spaces, we are in the familiar setting of differential calculus of several variables. As before, the charts need not be given explicitly if the property is independent of the choice of charts. We say instead that f has the property at p *relative to some (hence every) choice of charts* or just *relative to charts*, or, still more concisely, that f has this property at p. In particular:

Definition. A continuous map $f : M \to N$ between manifolds is called ***differentiable at*** $p \in M$ if it is differentiable at p relative to charts, and ***differentiable*** if it is differentiable at *every* $p \in M$. If f is bijective and both f and f^{-1} are differentiable, f is said to be a ***diffeomorphism***.

1.3 The Rank

The Jacobian matrix of the downstairs map is *not* independent of the choice of charts; after all, it transforms according to the chain rule under changes of charts. But the *rank* of the Jacobian matrix stays the same since the transition maps are diffeomorphisms, and we can make the following definition.

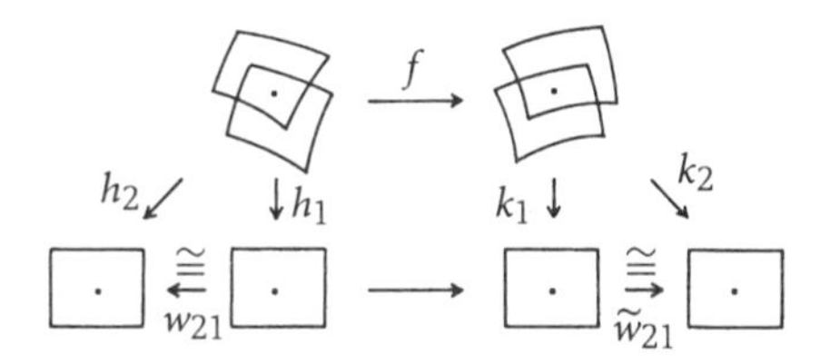

Figure 1.7. Why f has the same rank at p relative to (h_1, k_1) and to (h_2, k_2)

Definition and Remark. If $f : M \to N$ is differentiable at p, the rank of the Jacobian matrix relative to charts is called the ***rank*** of f at p and denoted by $\text{rk}_p f$. Note that if $f : M \to N$ is a diffeomorphism, the Jacobian matrix must be a square matrix of full rank. In particular, we see that diffeomorphic manifolds must have the same dimension.

As you know from the differential calculus of several variables, the rank governs basic properties of the *local* behavior of differentiable maps. The relevant theorems of differential calculus carry over directly to maps between manifolds because we can apply them to the downstairs maps. In this setting, the inverse function theorem reads:

Inverse Function Theorem. *If $f : M \to N$ is a differentiable map between two manifolds of the same dimension n and $p \in M$ is a point with $\mathrm{rk}_p f = n$, then f is a local diffeomorphism at p.*

Many fundamental local results in differential calculus follow as corollaries of the inverse function theorem. The (apparently more general) regular point theorem is one example.

Regular Point Theorem. *If $f : M \to N$ is a differentiable map between two manifolds and $p \in M$ is a* ***regular point*** *of f (that is, $\mathrm{rk}_p f = \dim N$), then f is locally at p (relative to suitable charts) the canonical projection.*

Spelled out in detail, this means that there are charts (U, h) around p and (V, k) about $f(p)$ such that $f(U) \subset V$ and the downstairs map $k \circ f \circ h^{-1} : U' \to V'$ is given (for instance) by

$$(x_1, \ldots, x_s, x_{s+1}, \ldots, x_{s+n}) \longmapsto (x_{s+1}, \ldots, x_{s+n}),$$

where $s + n$ and n denote the dimensions of M and N.

Another consequence of the inverse function theorem is the even more general *rank theorem*; see, for example, [BJ], p. 45:

Rank Theorem. *If the differentiable map $f : M \to N$ has constant rank r in a neighborhood of $p \in M$, then locally at p (relative to suitable charts) it is of the form*

$$\begin{aligned} \mathbb{R}^r \times \mathbb{R}^s &\longrightarrow \mathbb{R}^r \times \mathbb{R}^n, \\ (x, y) &\longmapsto (x, 0), \end{aligned}$$

where $r + s$ and $r + n$ are the dimensions of M and N.

1.4 Submanifolds

The regular point theorem makes an important statement about the preimage $f^{-1}(q)$ of a point $q \in N$, provided that the elements $p \in f^{-1}(q)$ are all regular. Such points q are called *regular values*.

Terminology. If $f : M \to N$ is a differentiable map, then the points $p \in M$ that are *not* regular are called ***critical***, or ***singular***, ***points*** of f, their images under f ***critical***, or ***singular***, ***values*** of f, and all the remaining points of N ***regular values*** of f.

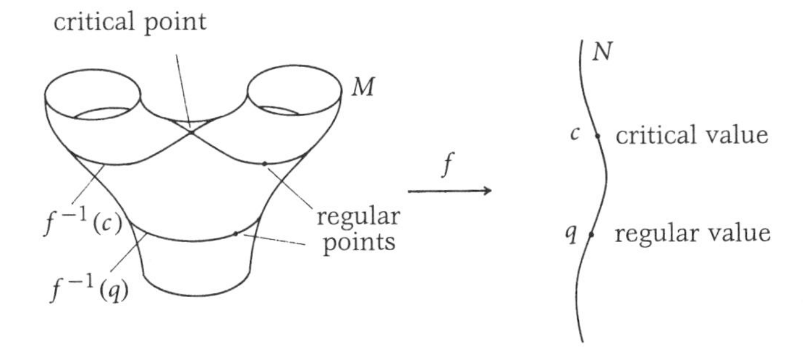

Figure 1.8. Regular and critical points and values

Observe that this fixes the convention of calling a point $q \in N$ a regular value if $f^{-1}(q)$ is empty, although such a q isn't even a "value" of f.

If M and N are manifolds with $\dim M = n+s$ and $\dim N = n$, and if $q \in N$ is a regular value of a differentiable map $f : M \to N$, then around every point p in the preimage $M_0 := f^{-1}(q)$ there is a chart (U, h) on M with the property

$$h(U \cap M_0) = \mathbb{R}^s \cap h(U),$$

where, as usual, we think of $\mathbb{R}^s \subset \mathbb{R}^{s+n}$ as $\mathbb{R}^s \times \{0\} \subset \mathbb{R}^s \times \mathbb{R}^n$. This is true because there is no problem in requiring that the chart (V, k) given by the regular point theorem satisfy $k(q) = 0$, and the corresponding (U, h) then does what we want.

Thus the entire subset M_0 of M, relative to suitable charts, lies in M as $\mathbb{R}^s$ does in $\mathbb{R}^{s+n}$, and is therefore called an s-dimensional *submanifold* of M. More precisely:

Definition. Let M be an n-dimensional manifold. A subspace $M_0 \subset M$ is a ***k-dimensional submanifold*** if around every point of M_0 there is a chart (U, h) on M with $h(U \cap M_0) = \mathbb{R}^k \cap h(U)$. Such a chart will be called a ***submanifold chart*** or, informally, a ***flattener*** for M_0 in M. The number $n - k$ is the ***codimension*** of M_0 in M.

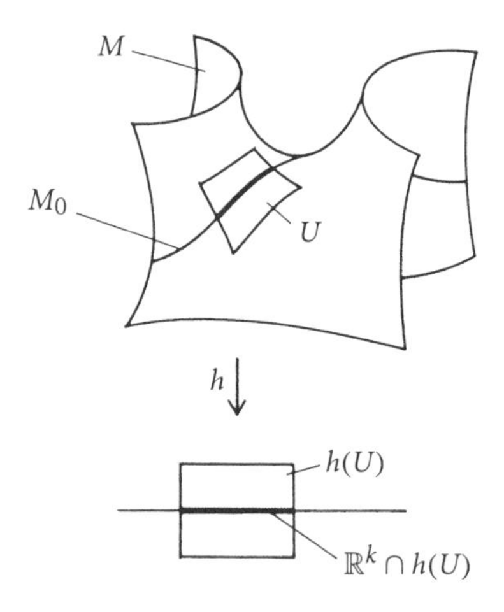

Figure 1.9. Flattener

Of course, M_0 isn't called a submanifold for nothing: The set $\mathfrak{A}_0$ of charts $(U \cap M_0, h|U \cap M_0)$ we get from the flatteners is obviously a k-dimensional differentiable atlas on M_0; it generates a differentiable structure $\mathcal{D}(\mathfrak{A}_0) =: \mathcal{D}|M_0$. Since the properties of being second countable and Hausdorff are inherited by subspaces, $(M_0, \mathcal{D}|M_0)$ is a k-dimensional differentiable manifold, and this is how we'll think of M_0 from now on. In the two extreme cases $k = 0$ and $k = n$, the submanifold condition reduces to a purely topological constraint: the zero-dimensional submanifolds of M are exactly the *discrete* subsets of M, and the zero-codimensional submanifolds of M are its *open* subsets.

What we said about the preimage of a regular value can now be put concisely as follows:

Regular Value Theorem. *If $q \in N$ is a regular value of a differentiable map $f : M \to N$, then its preimage $f^{-1}(q) \subset M$ is a submanifold whose codimension is equal to the dimension of N.*

1.5 Examples of Manifolds

Do manifolds really exist? Strictly speaking, apart from the every-dimensional *empty* manifold, I haven't yet given a single example.

To give manifolds straight from the definition, without resorting to any other tools, requires describing a second-countable Hausdorff space M and a differentiable structure $\mathcal{D}$ on M. Of course, only one (preferably *small*) differentiable atlas $\mathfrak{A}$ on M need be given explicitly in order to define $\mathcal{D}$ as the maximal atlas $\mathcal{D}(\mathfrak{A})$ containing $\mathfrak{A}$. The easiest manifold to obtain in this way is the local model for all n-dimensional manifolds, $\mathbb{R}^n$, which we naturally take to be the manifold

$$(\mathbb{R}^n, \mathcal{D}(\{\mathrm{Id}_{\mathbb{R}^n}\})).$$

And this is the *only* manifold I'll give straight from the definition! In real life, you hardly ever come across manifolds this way. Let me explain this by a comparison from calculus.

A real function of a real variable is called *continuous* at x_0 if for every $\varepsilon > 0$ there is a $\delta > 0$ such that, etc. From this, it is easy to see that constant functions are continuous (δ arbitrary) and the identity function is continuous (for example, $\delta := \varepsilon$). But if you have to justify why the function defined by $f(x) := \arctan(x+\sqrt{x^4 + e^{\cosh x}})$, or something like it, is continuous, do you start with an arbitrary $\varepsilon > 0$ and look for a $\delta > 0$ such that ...? No. Instead, from the theory you recall that there are *processes that produce continuous functions*—for instance, sums, products, quotients, uniformly convergent series, composition, and inverses (on intervals of monotonicity) of continuous functions are continuous—and you immediately see that the function above comes from applying such processes to the constant function and the identity function.

Instead of explicitly setting out the defining properties and characteristics of mathematical objects, one can often get by with recalling *where they come from* and *how they are formed*. There are *processes that produce manifolds*, and the regular value theorem, for instance, is a wellspring. The map $f : \mathbb{R}^{n+1} \to \mathbb{R}$ given by $f(x) := \|x\|^2$ has rank 1 everywhere except at $x = 0$. In particular, $1 \in \mathbb{R}$ is a regular value, and its preimage $f^{-1}(1)$, the ***n*-sphere** $S^n := \{x \in \mathbb{R}^{n+1} : \|x\| = 1\}$, is therefore an n-dimensional submanifold of $\mathbb{R}^{n+1}$. The map $f : \mathbb{R}^3 \to \mathbb{R}$, $x \mapsto x_1^2 + x_2^2 - x_3^2$, is also singular only at $x = 0$; hence every $c \neq 0$ in $\mathbb{R}$ is a regular value of f, and the ***hyperboloid*** $f^{-1}(c)$ is a two-dimensional submanifold of $\mathbb{R}^3$ (a "surface in space").

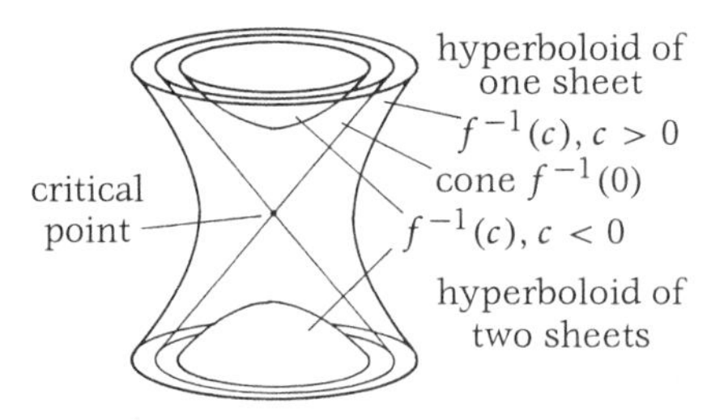

Figure 1.10. Hyperboloids as submanifolds, according to the regular value theorem

I would also like to mention a third application of the regular value theorem, this one a bit more interesting. This time the two manifolds M and N will be finite-dimensional vector spaces. To be precise, let $n \geq 1$; let $M := M(n \times n, \mathbb{R}) \cong \mathbb{R}^{n^2}$, the space of real $n \times n$-matrices; and let $N := S(n \times n, \mathbb{R}) \cong \mathbb{R}^{\frac{1}{2}n(n+1)}$, the subspace of *symmetric* matrices. If $A \in M(n \times n, \mathbb{R})$, we denote its transpose by ${}^tA$. Let I be the $n \times n$ identity matrix. Recall that a matrix A is called *orthogonal* if ${}^tA \cdot A = I$.

Lemma. *The identity matrix I is a regular value of the map*

$$\begin{aligned} f: M(n \times n, \mathbb{R}) &\longrightarrow S(n \times n, \mathbb{R}), \\ A &\longmapsto {}^tA \cdot A. \end{aligned}$$

Hence the ***orthogonal group***

$$O(n) := f^{-1}(I)$$

is a $\frac{1}{2}n(n-1)$-dimensional submanifold of $M(n \times n, \mathbb{R})$.

Proof. We must show that f is regular at $A \in O(n)$. Rather than finding its rank by explicitly computing a $\frac{1}{2}n(n+1) \times n^2$ Jacobian matrix, we recall the relationship between the Jacobian matrix and the directional derivative in differential calculus: In general,

$$J_f(p) \cdot v = \frac{d}{d\lambda}\bigg|_0 f(p + \lambda v).$$

Thus it suffices to prove that for every $A \in O(n)$ and every $B \in S(n \times n, \mathbb{R})$, there is a matrix $X \in M(n \times n, \mathbb{R})$ such that

$$\frac{d}{d\lambda}\bigg|_0 {}^t(A + \lambda X) \cdot (A + \lambda X) = B,$$

i.e. $J_f(A)X = B$. This will show that the Jacobian matrix of f relative to linear charts is a surjective map $\mathbb{R}^{n^2} \to \mathbb{R}^{\frac{1}{2}n(n+1)}$, hence that f is of full rank $\frac{1}{2}n(n+1)$ at A. So all we have to do is find, for every symmetric matrix B, a matrix X with

$${}^tX \cdot A + {}^tA \cdot X = B.$$

Since B is symmetric and ${}^tX \cdot A = {}^t({}^tA \cdot X)$, it suffices to find X such that

$${}^tA \cdot X = \tfrac{1}{2}B.$$

But we can do this not just for orthogonal A but for any invertible A, by setting $X := \frac{1}{2}{}^tA^{-1}B$. $\square$

Observe that this also proves that the ***special orthogonal group***

$$SO(n) := \{A \in O(n) | \det A = +1\}$$

is a $\frac{1}{2}n(n-1)$-dimensional submanifold of $M(n \times n, \mathbb{R})$, since $SO(n)$ is open in $O(n)$. Applying the regular value theorem in a completely similar way shows that other "matrix groups," such as $U(n)$ or $SU(n)$, are submanifolds of vector spaces of matrices.

In linear algebra, one studies *linear* systems of equations $A \cdot x = b$. The solution set of such a system is nothing but the preimage $A^{-1}(b)$ of the value b under the linear map A. Now the preimages $f^{-1}(q)$ of differentiable maps are just the solution sets of *nonlinear* systems of equations and are submanifolds when q is regular, so one can do analysis on them. This is one reason for studying manifolds.

1.6 Sums, Products, and Quotients of Manifolds

In this section we discuss three more processes that yield manifolds, namely taking *sums, products,* and *quotients*. The most primitive of these is summation, the simple juxtaposition of manifolds by disjoint union.

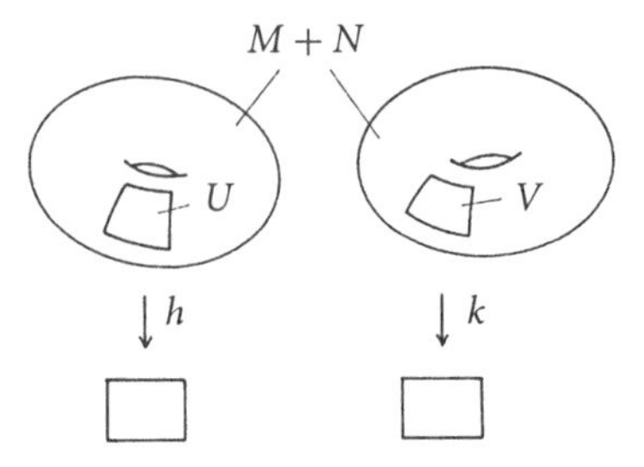

Figure 1.11. Charts for the summands are also charts for the sum; the atlas $\mathfrak{A} + \mathfrak{B}$ is differentiable because no new transition maps have been added.

Note. *If M and N are n-dimensional manifolds, then, in a canonical way, so is their* ***sum****, or disjoint union, $M + N$.*

If $\mathfrak{A}$ and $\mathfrak{B}$ are atlases on M and N, respectively, their disjoint union $\mathfrak{A} \dot{\cup} \mathfrak{B} =: \mathfrak{A} + \mathfrak{B}$ is obviously an atlas on $M+N$, and if we wanted to state the note above a bit more formally we would have to give the differentiable structure on $M + N$ as $\mathcal{D}(\mathcal{D}_1 + \mathcal{D}_2)$, where $\mathcal{D}_1$ and $\mathcal{D}_2$ are the structures on M and N, respectively. Then $\mathcal{D}(\mathcal{D}(\mathfrak{A}) + \mathcal{D}(\mathfrak{B})) = \mathcal{D}(\mathfrak{A} + \mathfrak{B})$ might also be worth mentioning.

Of course, we can deal similarly with several or even countably many summands M_i, $i = 1, 2, \ldots$ and take their sum, or disjoint union,

$$\coprod_{i=1}^{\infty} M_i,$$

but this won't work for uncountably many summands because the second axiom of countability still has to be satisfied.

We often have to take the *product* of two manifolds. Topologically, of course, this just means taking the Cartesian product, and the differentiable structure is obtained from the products of charts on the factors.

Note. *The* ***product*** *$M \times N$ of a k-dimensional manifold with an n-dimensional manifold is canonically a $(k + n)$-dimensional manifold.*

We may safely permit ourselves the notation

$$\mathfrak{A} \times \mathfrak{B} := \{(U \times V, h \times k) \ : \ (U, h) \in \mathfrak{A}, \ (V, k) \in \mathfrak{B}\}$$

for the ***product atlas*** because the product of the charts,

$$\begin{array}{c} U \times V \\ \cong \Big\downarrow h \times k \\ U' \times V' \underset{\text{open}}{\subset} \mathbb{R}^k \times \mathbb{R}^n = \mathbb{R}^{k+n}, \end{array}$$

contains the same information as the pair (h, k) unless one of the two charts is empty. In this notation, the differentiable structure on $M \times N$ intended in the note is of course $\mathcal{D} := \mathcal{D}(\mathcal{D}_1 \times \mathcal{D}_2)$, where $\mathcal{D}_1 = \mathcal{D}(\mathfrak{A})$ and $\mathcal{D}_2 = \mathcal{D}(\mathfrak{B})$ are the structures on M and N, respectively, and it is easy to see that then we also have $\mathcal{D} = \mathcal{D}(\mathfrak{A} \times \mathfrak{B})$.

Perhaps the simplest nontrivial example of a product manifold is the ***torus*** $T^2 := S^1 \times S^1$, which we often use as an illustration. If we think of $\mathbb{R}^2$ as $\mathbb{C}$, then $S^1 = \{z \in \mathbb{C} : |z| = 1\}$, so we could find $S^1 \times S^1$ in $\mathbb{C}^2$. But since this is hard to draw, we substitute a submanifold of $\mathbb{R}^3$ diffeomorphic to $S^1 \times S^1$.

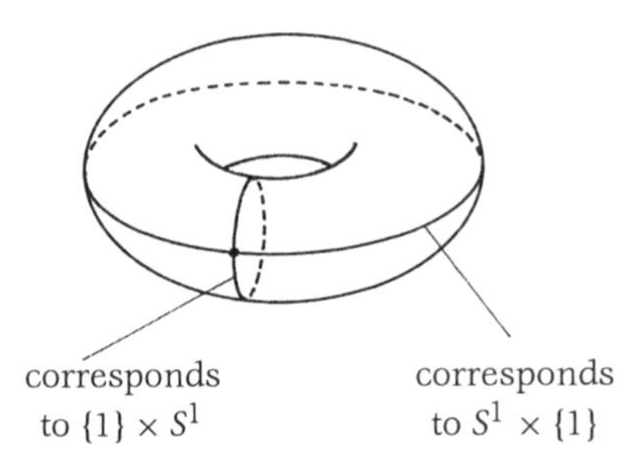

Figure 1.12. A torus represented in $\mathbb{R}^3$

The subject of quotient manifolds is more subtle, and for now we can only take a first step in that direction.

Let X be a topological space and $\sim$ an equivalence relation on X, and let $X/\sim$ denote the set of equivalence classes. If

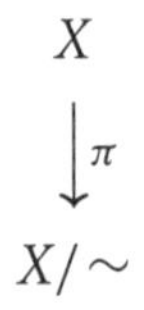

$$X \xrightarrow{\pi} X/\sim$$

is the canonical projection that assigns to every $x \in X$ its equivalence class, then $U \subset X/\sim$ is called ***open in the quotient topology*** if $\pi^{-1}(U)$ is open in X, and $X/\sim$, endowed with the quotient topology, is called the ***quotient space*** of X under $\sim$.

So much for recalling a topological notion (see, for example, Chapter III of [J:*Top*], pp. 31–33 in particular). Now, if M is a manifold and $\sim$ an equivalence relation on it, then $M/\sim$ is—a long way from being a manifold, and often not even a Hausdorff space. We consider here what is in some sense the simplest case in which $M/\sim$ is a manifold.

Lemma. *Let M be an n-dimensional manifold, $\tau : M \to M$ a* ***fixed-point-free involution*** *(i.e. a differentiable map with $\tau \circ \tau = \mathrm{Id}_M$ and $\tau(x) \neq x$ for all $x \in M$), and M/τ the quotient space of M under the equivalence relation $x \sim \tau(x)$. Then M/τ is also an n-dimensional manifold in a canonical way: its differentiable structure is the only one for which*

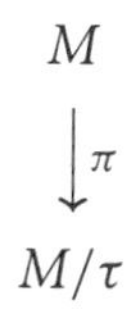

is a local diffeomorphism everywhere.

Proof. Of course there can be at most *one* such structure, for the identity on M/τ with respect to two such structures would be a local diffeomorphism, hence in fact a diffeomorphism (see Exercise 1.2):

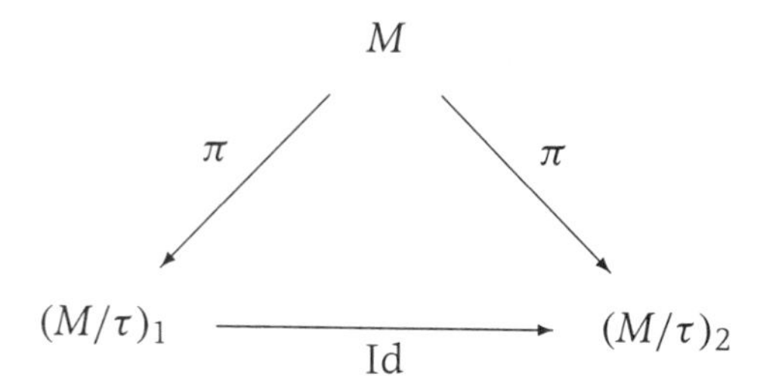

To prove that M/τ is a Hausdorff space, we consider two points $\pi(p) \neq \pi(q) \in M/\tau$. Since M is a Hausdorff space, we can choose open neighborhoods U and V of p and q, respectively, so small that $U \cap V = \emptyset$ and $U \cap \tau(V) = \emptyset$. Then $\pi(U)$ and $\pi(V)$ are separating neighborhoods of $\pi(p)$ and $\pi(q)$.

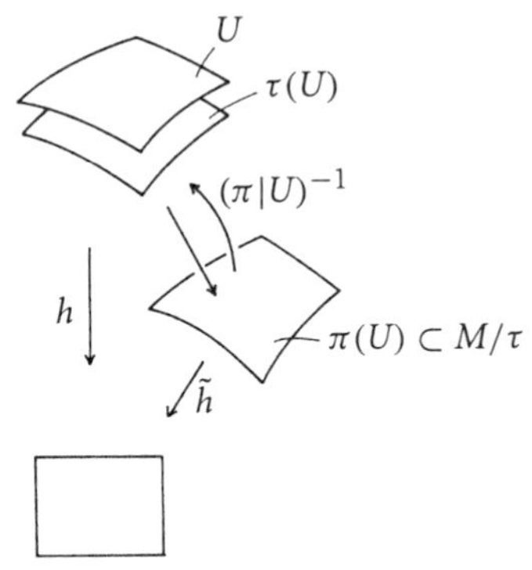

Figure 1.13. Charts on the quotient manifold M/τ

Moreover, if $\{U_i\}_{i\in\mathbb{N}}$ is a countable basis for M, then $\{\pi(U_i)\}_{i\in\mathbb{N}}$ is a countable basis for N. We have not yet used the fact that τ is fixed-point free. We use it now in defining a subset U of M to be *small* if $U \cap \tau(U) = \emptyset$ and convincing ourselves that M is "locally small"; that is, every neighborhood of a point contains a small neighborhood. If $U \subset M$ is a small open set, then $\pi|U : U \xrightarrow{\cong} \pi(U)$ is a homeomorphism, so every small chart (U, h) on M defines a chart $(\pi(U), \widetilde{h})$ on M/τ.

The small charts form an atlas $\mathfrak{A}$ on M, and

$$\widetilde{\mathfrak{A}} := \{(\pi(U), \widetilde{h}) \; : \; (U, h) \in \mathfrak{A}\}$$

is an atlas on M/τ. The corresponding differentiable structure $\mathcal{D}(\widetilde{\mathfrak{A}})$ has the desired property. □

Example. The quotient manifold

$$\mathbb{RP}^n := S^n/-\mathrm{Id}$$

of the n-sphere under the antipodal involution $x \mapsto -x$ is n-dimensional ***real projective space***.

Perhaps I should say that this is real projective space *as a differential-topological object*. From the algebraic viewpoint, resorting to the sphere to define projective space is misleading. If V is any vector space over an arbitrary field $\mathbb{K}$, the corresponding projective space $\mathbb{KP}(V)$ can be defined as the set of one-dimensional subspaces of V. In particular, $\mathbb{KP}^n := \mathbb{KP}(\mathbb{K}^{n+1})$ can be defined without any need for a norm on V or $\mathbb{K}^{n+1}$. But for $\mathbb{K} = \mathbb{R}$ it is obvious that $\mathbb{RP}(\mathbb{R}^{n+1}) = S^n/-\mathrm{Id}$, and the quotient map $S^n \to \mathbb{RP}^n$ is very useful for looking at $\mathbb{RP}^n$ differential-topologically.

It is also easy to give an atlas for $\mathbb{RP}^n$ directly: if the points of projective space are described in "homogeneous coordinates" by $[x_0 : \ldots : x_n] \in \mathbb{RP}^n$ for $(x_0, \ldots, x_n) \in \mathbb{R}^{n+1}\setminus\{0\}$, then an atlas with $n+1$ charts is defined by $U_i := \{[x] \; : \; x_i \neq 0\}$ and $h_i[x] := (x_0/x_i, \ldots, \widehat{\imath}, \ldots, x_n/x_i)$ for $i = 0, \ldots, n$.

1.7 Will Submanifolds of Euclidean Spaces Do?

I would like to end this chapter by pointing out a particular aspect of taking quotients.

If we start with $\mathbb{R}^n$ and its open submanifolds as the simplest examples and create new manifolds by taking regular preimages, sums, and products, we always get submanifolds of Euclidean spaces back again. Nothing completely new happens until we take quotients. Then, for instance, we get a "surface" $\mathbb{RP}^2 = S^2/\sim$ that no longer lies in the space $\mathbb{R}^3$. This, much more than, say, the sphere S^2, which we can also picture as a geometric locus in $\mathbb{R}^3$, makes it obvious that we need a mathematical formulation of the notion of an "intrinsic surface" (more generally, what we need is precisely the notion of a manifold).

This is a fine observation as far as it goes, but I don't want to conceal that there is a classical theorem in differential topology, the *Whitney embedding theorem*, that seems to point in the opposite direction. A map $f : M \to N$ is called an ***embedding*** if $f(M) \subset N$ is a submanifold and $f : M \xrightarrow{\cong} f(M)$ is a diffeomorphism. Now, the ***Whitney embedding theorem*** (see, for instance, [BJ], p. 71) says that *every* n-dimensional manifold can be embedded in $\mathbb{R}^{2n+1}$, and even that there exists an embedding with closed image. *Thus every manifold is diffeomorphic to a closed submanifold of some* $\mathbb{R}^N$! Do we still really need "abstract" manifolds?

Well, the embeddability of manifolds in $\mathbb{R}^N$ is one of several interesting properties of these objects, and is sometimes useful in proofs and constructions. But, as you know, the mere *existence* of a thing doesn't mean that the thing is within easy reach or given canonically. So we shouldn't expect manifolds, as we encounter them in nature—as quotient manifolds, for instance—to be carrying an embedding into some $\mathbb{R}^N$ in their luggage. If, in the deceptive hope of convenience, we restricted our further development of differential-topological ideas to submanifolds of $\mathbb{R}^N$, then in every application to a "natural manifold" we would have to start by embedding it (which can be quite tedious in concrete cases), then keep the dependence of the concepts and constructions on the *choice* of the embedding under control (for there is usually no canonical embedding), and at the end we wouldn't even be rewarded for all our efforts, since submanifolds of $\mathbb{R}^n$ are by no means easier to handle. After all, how they lie in $\mathbb{R}^n$ has to be described somehow by equations and conditions, and in the coordinates of the ambient space the formulas—those for integration on manifolds, for instance—actually become *messier* instead of simpler.

This is why, in the next chapter, we painstakingly develop the key idea of the tangent space for arbitrary manifolds that don't necessarily lie in any $\mathbb{R}^N$.

1.8 Test

(1) Is every n-dimensional chart simultaneously an m-dimensional chart for all $m \geq n$?

- ☐ Yes.
- ☐ This is a matter of opinion, and depends on whether or not you want to distinguish between $\mathbb{R}^n$ and $\mathbb{R}^n \times \{0\} \subset \mathbb{R}^m$ in this context.
- ☐ No, because U' isn't open in $\mathbb{R}^m$ if $U \neq \emptyset$ and $m > n$.

(2) Does the differentiable structure $\mathcal{D}$ on an n-dimensional manifold $(M, \mathcal{D})$ consist exactly of all the diffeomorphisms between open subsets U of M and open subsets U' of $\mathbb{R}^n$?

- ☐ Yes.
- ☐ No, because the charts don't have to be diffeomorphisms (just homeomorphisms).
- ☐ No, because in general there are many more such diffeomorphisms than elements of $\mathcal{D}$.

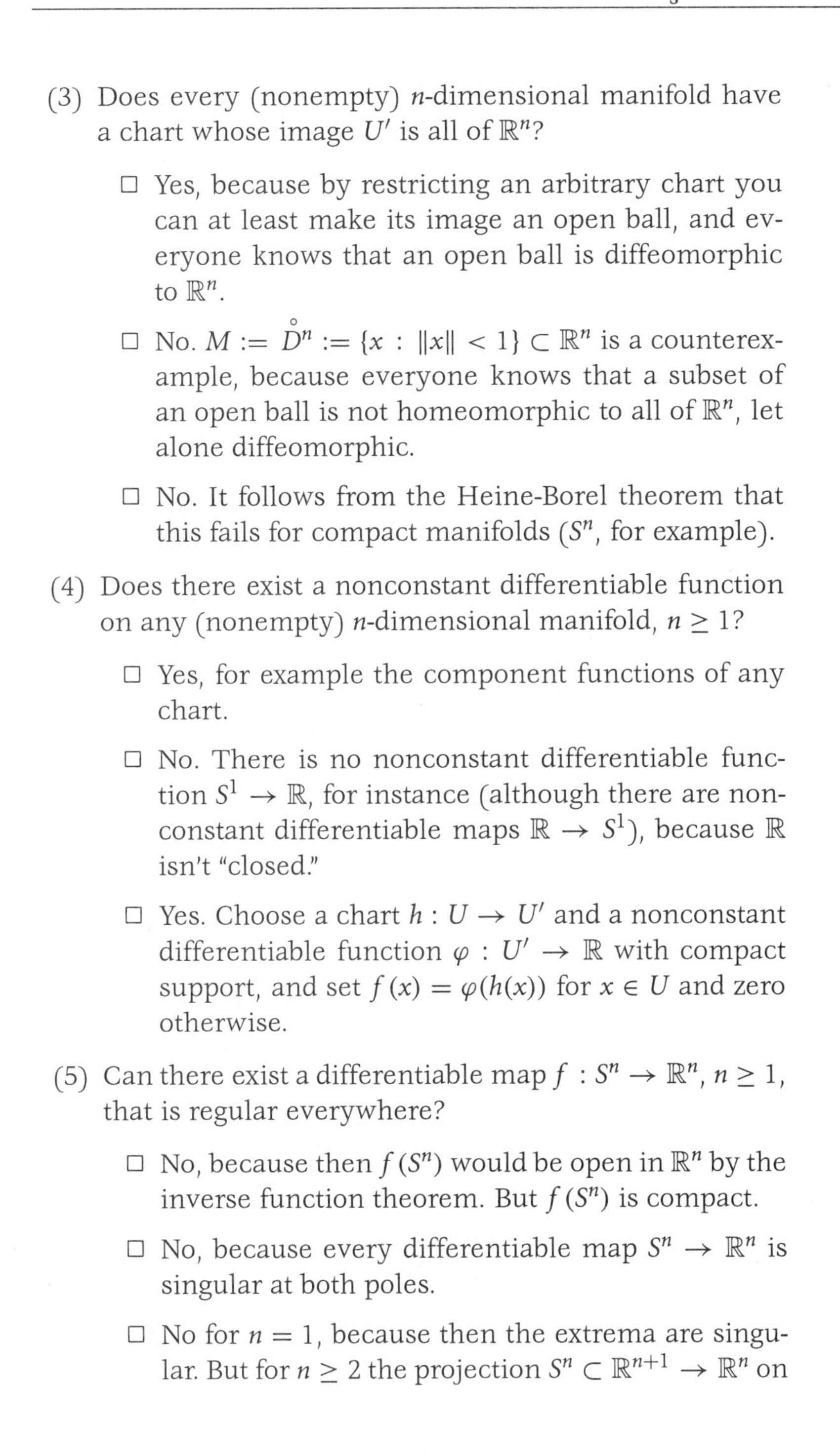

(3) Does every (nonempty) n-dimensional manifold have a chart whose image U' is all of $\mathbb{R}^n$?

□ Yes, because by restricting an arbitrary chart you can at least make its image an open ball, and everyone knows that an open ball is diffeomorphic to $\mathbb{R}^n$.

□ No. $M := \overset{\circ}{D}{}^n := \{x : \|x\| < 1\} \subset \mathbb{R}^n$ is a counterexample, because everyone knows that a subset of an open ball is not homeomorphic to all of $\mathbb{R}^n$, let alone diffeomorphic.

□ No. It follows from the Heine-Borel theorem that this fails for compact manifolds (S^n, for example).

(4) Does there exist a nonconstant differentiable function on any (nonempty) n-dimensional manifold, $n \geq 1$?

□ Yes, for example the component functions of any chart.

□ No. There is no nonconstant differentiable function $S^1 \to \mathbb{R}$, for instance (although there are nonconstant differentiable maps $\mathbb{R} \to S^1$), because $\mathbb{R}$ isn't "closed."

□ Yes. Choose a chart $h : U \to U'$ and a nonconstant differentiable function $\varphi : U' \to \mathbb{R}$ with compact support, and set $f(x) = \varphi(h(x))$ for $x \in U$ and zero otherwise.

(5) Can there exist a differentiable map $f : S^n \to \mathbb{R}^n$, $n \geq 1$, that is regular everywhere?

□ No, because then $f(S^n)$ would be open in $\mathbb{R}^n$ by the inverse function theorem. But $f(S^n)$ is compact.

□ No, because every differentiable map $S^n \to \mathbb{R}^n$ is singular at both poles.

□ No for $n = 1$, because then the extrema are singular. But for $n \geq 2$ the projection $S^n \subset \mathbb{R}^{n+1} \to \mathbb{R}^n$ on the first n coordinates, for instance, has the desired property.

(6) Which of the following three sketches could, in the eyes of a sympathetic reader, represent a two-dimensional submanifold of $\mathbb{R}^3$?

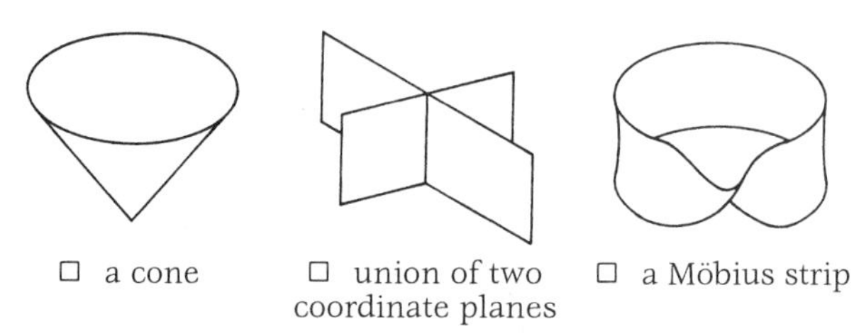

☐ a cone ☐ union of two coordinate planes ☐ a Möbius strip

Figure 1.14.

(7) Figure 1.15 shows a map $(x, y, z) \mapsto (x, y)$ from a two-dimensional submanifold $M \subset \mathbb{R}^3$ to the plane. What ranks occur?

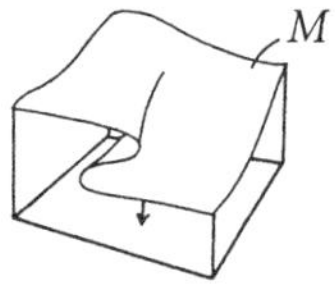

- ☐ Only rank 2.
- ☐ Only ranks 1 and 2.
- ☐ All three ranks 0, 1, and 2.

M

Figure 1.15.

(8) Is there a surjective map $f : \mathbb{R}^2 \to S^1 \times S^1$ that is regular everywhere?

- ☐ No. Since $S^1 \times S^1$ is compact and $\mathbb{R}^2$ is not, the inverse function theorem gives a contradiction.
- ☐ Yes. An example is $f(x, y) := (e^{ix}, e^{iy})$.
- ☐ Yes, because for connected two-dimensional M there's always such a map $f : \mathbb{R}^2 \to M$. (Picture a long wide brushstroke.)

(9) The following sketches show maps from a closed rectangle to $\mathbb{R}^3$. Which of them could define an embedding of the interior of the rectangle?

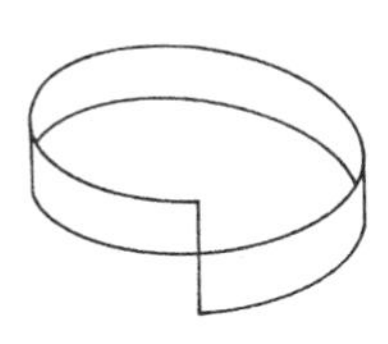
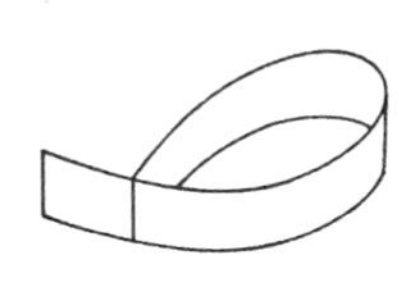
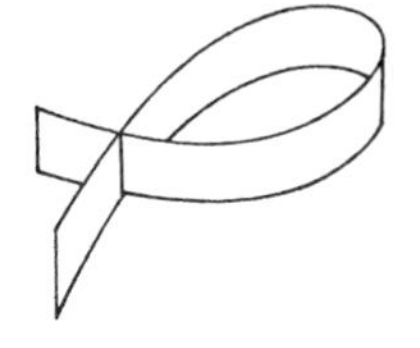
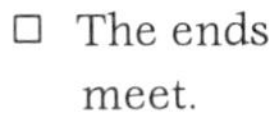

□ The ends meet. □ An end meets the interior. □ Self-intersection

Figure 1.16.

(10) Must the quotient $M/\sim$ of a manifold be Hausdorff if every equivalence class consists of exactly two points?

□ Yes. In fact, this holds whenever the equivalence classes are finite.

□ Yes, and it's crucial here that no 1-point class is allowed. Otherwise, in $\{0, 1\} \times \mathbb{R}$ identify $(0, x)$ and $(1, x)$ for each $x \neq 0$; then $(0, 0)$ and $(1, 0)$ can't be separated.

□ No. For example, let $M = S^1 \subset \mathbb{C}$. Set $1 \sim i$ and $-1 \sim -i$, and set $z \sim \bar{z}$ otherwise.

1.9 Exercises

Exercise 1.1. Prove that every manifold has a countable atlas.

Exercise 1.2. Let $\mathcal{D}_1$ and $\mathcal{D}_2$ be differentiable structures on the same second-countable Hausdorff space M. Prove that the identity on M is a *diffeomorphism* between $\mathcal{D}_1$ and $\mathcal{D}_2$ if and only if $\mathcal{D}_1 = \mathcal{D}_2$.

Exercise 1.3. State more precisely and prove: Every n-dimensional real vector space is an n-dimensional differentiable manifold in a canonical way.

Exercise 1.4. Let M be a differentiable manifold, $\dim M \geq 1$, and let $p \in M$. Prove that $M \setminus \{p\}$ is not compact.

EXERCISE 1.5. Prove that $S^n \times S^k$ is diffeomorphic to a submanifold of $\mathbb{R}^{n+k+1}$. (Hint: First show that $S^n \times \mathbb{R}$ and $\mathbb{R}^{n+1} \setminus \{0\}$ are diffeomorphic.)

EXERCISE 1.6. Let M be an n-dimensional manifold and let X and Y be two disjoint closed k-dimensional submanifolds of M. Show that $X \cup Y$ is also a submanifold of M. Why can't you just omit the hypothesis that X and Y are closed?

EXERCISE 1.7. Let $Q : \mathbb{R}^n \to \mathbb{R}$ be a nondegenerate quadratic form on $\mathbb{R}^n$. Show that the group

$$O(Q) := \{A \in GL(n, \mathbb{R}) \ : \ Q \circ A = Q\}$$

is a submanifold (of what dimension?) of $GL(n, \mathbb{R})$.

EXERCISE 1.8. Show that every manifold is the sum of its path components.

1.10 Hints for the Exercises

FOR EXERCISE 1.1. At least the topology of M has a countable basis $(\Omega_i)_{i \in \mathbb{N}}$. Is every Ω_i contained in a chart domain for some chart (U_i, h_i) of the differentiable structure $\mathcal{D}$? If so, would $\{(U_i, h_i) : i \in \mathbb{N}\}$ be an atlas? This takes some thought. The answer to the first question, for instance, is *no* in general. Ω_i may be too "big." What can be done if it is?

FOR EXERCISE 1.2. This is an exercise in the definitions. No ideas are needed here. You "only" have to prove both directions $\Longrightarrow$ and $\Longleftarrow$ directly from the definitions.

FOR EXERCISE 1.3. Maybe you have no idea where to start with "state more precisely" and are muttering that I should have *formulated the problem more precisely*. The convenient phrase "in a canonical way" says something meaningful only if it's clear *what* way we're really talking about. An n-dimensional real vector space is, in any case, not an n-dimensional manifold according to the strict wording of the definition. That much is clear. The exercise must be about endowing V in some obvious way with a topology (how?) and a differentiable structure (how?) *that turn V into* a manifold. Of course, I could give these data precisely and just leave you to prove that the properties required by the definition of a manifold are satisfied. But this would take away most of the content of the exercise. The point is to practice giving a precise meaning to the expression "in a canonical way" by yourself. Mathematics just can't be done without it.

FOR EXERCISE 1.4. You can certainly give reasons why the ball minus the origin, $D^n \setminus \{0\}$, is not compact: the Heine-Borel theorem tells us so, for instance, or we can see immediately that the open cover by the sets $U_k := \{x : |x| > 1/k\}$ has no finite subcover, or you might refer to the fact that the sequence $(1/k)_{k=1,2,\dots}$ has no subsequence that converges in $D^n \setminus \{0\}$.

Can this situation be used somehow for the exercise, by means of a chart around p? Well, yes, somehow. But be careful: the assertion becomes false if we don't require M to be Hausdorff. So the Hausdorff property has to play a role in the proof!

FOR EXERCISE 1.5. Inherently, $S^n \times S^k \subset \mathbb{R}^{n+1} \times \mathbb{R}^{k+1} = \mathbb{R}^{n+k+2}$, one dimension too many. As an intermediate step, it might be a good idea to show that $S^n \times \mathbb{R} \cong \mathbb{R}^{n+1} \setminus \{0\}$. This suggests polar or spherical coordinates. But wouldn't this lead to $S^n \times \mathbb{R}_+ \cong \mathbb{R}^{n+1} \setminus \{0\}$, thus giving the factor $\mathbb{R}_+ := \{r \in \mathbb{R} : r > 0\}$ instead of $\mathbb{R}$? And what use would $S^n \times \mathbb{R} \cong \mathbb{R}^{n+1} \setminus \{0\}$ be for the exercise itself?

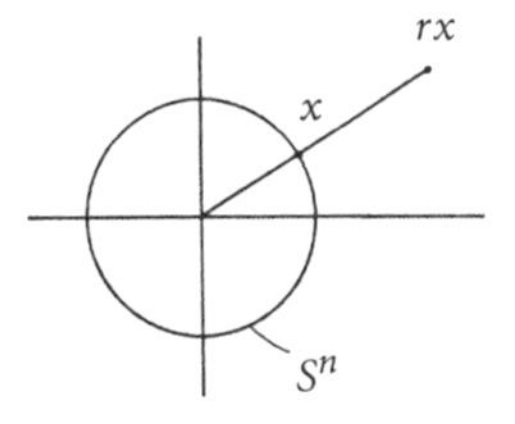

Figure 1.17.

FOR EXERCISE 1.6. The first part is a straightforward exercise in the definitions. For the additional question, you have to find a starting point by getting an intuitive idea of what's going on. With a bit of thought, you can find a counterexample even for $M = \mathbb{R}$ and $k = 0$. Well, that does it! Of course, it would be even better to prove that there are counterexamples for every n-dimensional $M \neq \emptyset$ and $0 \leq k \leq n-1$.

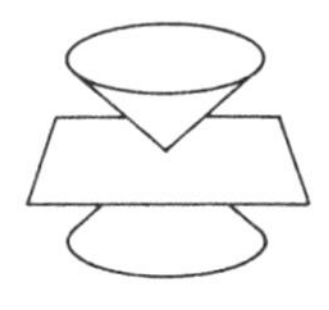

Figure 1.18.

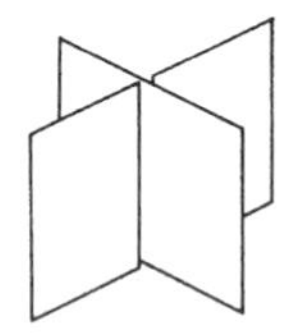

Figure 1.19.

Figures 1.18 and 1.19 should give you some idea how to proceed. The main problem, of course, is then the proof that a given subset of M really isn't a submanifold.

FOR EXERCISE 1.7. Matrix groups such as $O(Q)$ are important examples of *Lie groups*. For

$$Q(x) = x_0^2 - x_1^2 - x_2^2 - x_3^2$$

on $\mathbb{R}^4$, for instance, $O(Q)$ is the *Lorentz group*. You probably know from linear algebra (see Section 11.5 in [J:*LiA*], for example) that given a quadratic form Q on $\mathbb{R}^n$ there is a well-defined symmetric $n \times n$ matrix C such that $Q(x) = {}^t x \cdot C \cdot x$. That Q is nondegenerate means that C has rank n. If C is, in this sense, the matrix of the quadratic form Q, what is the matrix of $Q \circ A$? Now try to use the regular value theorem, as we did earlier (in Section 1.5) for $O(n)$.

FOR EXERCISE 1.8. If two points $a, b \in M$ are called equivalent, $a \sim b$, when they can be joined by a continuous path $\alpha : [0,1] \to M$, then the equivalence classes are called the *path components* of M. These path components are *open* (why?), and there can be only countably many (why?). Let there be $k \in \mathbb{N} \cup \{\infty\}$ of them, and let's think of them as being numbered, or "counted," $M_1, \ldots, M_k$ or $(M_i)_{i \in \mathbb{N}}$ (if $k = \infty$). You should now show that the canonical bijection

$$\coprod_{i=1}^{k} M_i \xrightarrow{\cong} M$$

(which bijection?) is a diffeomorphism. As far as content is concerned, this is a routine verification, but carrying it out will test whether your intuitive ideas about the sum can be replaced by watertight arguments.

2 CHAPTER The Tangent Space

2.1 Tangent Spaces in Euclidean Space

One of the basic ideas of differential calculus is to approximate differentiable maps by linear maps so as to reduce analytic (hard) problems to linear-algebraic (easy) problems whenever possible. Recall that locally at x, the linear approximation of a map $f : \mathbb{R}^n \to \mathbb{R}^k$ is the *differential* $df_x : \mathbb{R}^n \to \mathbb{R}^k$ of f at x. The differential is characterized by $f(x + v) = f(x) + df_x \cdot v + \varphi(v)$, where $\lim_{v\to 0} \varphi(v)/\|v\| = 0$, and given by the Jacobian matrix. But how can a differentiable map $f : M \to N$ between *manifolds* be characterized locally at $p \in M$ by a linear map?

Of course, we can always consider the differential $d(k \circ f \circ h^{-1})_x$ of the downstairs map. But this differential really does depend on the choice of charts—after all, it approximates $k \circ f \circ h^{-1}$, not f itself. If we want to define a differential for f that is *independent* of charts, we have some preliminary work to do: we have to begin by approximating the *manifolds* M and N locally at p and $f(p)$ "linearly," in other words by vector spaces. Only then can we define the

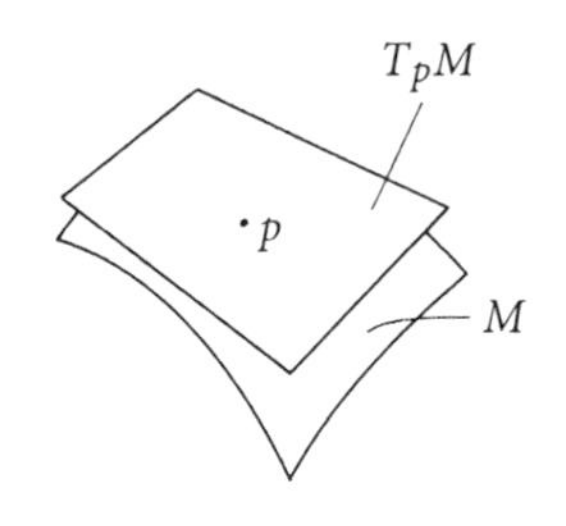

Figure 2.1. The tangent space T_pM

differential as a linear map

$$df_p : T_pM \to T_{f(p)}N$$

between these so-called *tangent spaces*. The purpose of the present chapter is to introduce these tangent spaces.

To orient ourselves, we first consider the submanifolds of Euclidean space $\mathbb{R}^N$. Here we have an obvious way of defining the tangent space—by analogy with the classical tangent plane to a surface in space:

Lemma and Definition. *If $M \subset \mathbb{R}^N$ is an n-dimensional submanifold, $p \in M$, and (U, h) is a chart on $\mathbb{R}^N$ around p that flattens M, then the vector subspace of $\mathbb{R}^N$ defined by*

$$T_p^{\text{sub}}M := (dh_p)^{-1}(\mathbb{R}^n \times \{0\})$$

is independent of the choice of charts. It is called the (submanifold) ***tangent space*** *of M at the point p.*

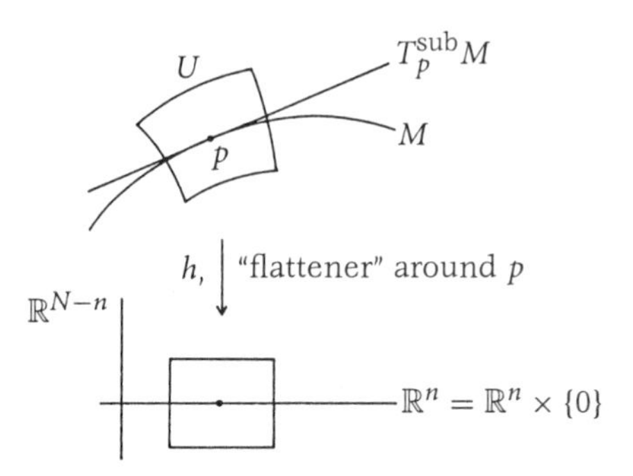

Figure 2.2. The tangent space to a submanifold of $\mathbb{R}^N$

PROOF. The transition map w between two flatteners (U, h) and $(V, \widetilde{h})$ around p has to take $h(U \cap V) \cap (\mathbb{R}^n \times \{0\})$ onto $\widetilde{h}(U \cap V) \cap (\mathbb{R}^n \times \{0\})$, so its differential at $h(p)$ maps $\mathbb{R}^n \times \{0\}$ onto $\mathbb{R}^n \times \{0\}$. The assertion follows because $(d\widetilde{h}_p)^{-1} = (dh_p)^{-1} \circ (dw_{h(p)})^{-1}$. Hence $T_p^{\text{sub}}M$ is well defined. □

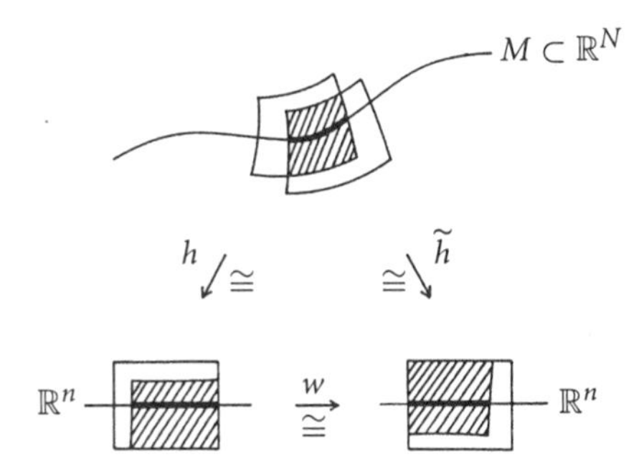

Figure 2.3. The transition map between two flatteners

It may not be completely unnecessary to point out that $T_p^{\text{sub}}M \subset \mathbb{R}^N$ is therefore really a *subspace* of $\mathbb{R}^N$, and in particular contains the zero vector $0 \in \mathbb{R}^N$. It's only by drawing pictures that we tend to shift it, by translation to p, to the place where our geometric intuition wants to see it. But we mustn't forget that its vector space structure is then the one that has the zero vector at the point p. This shouldn't lead to misunderstandings, though, any more than "attaching" the velocity vector $\dot{\alpha}(t)$ of a plane curve to the appropriate point $\alpha(t)$.

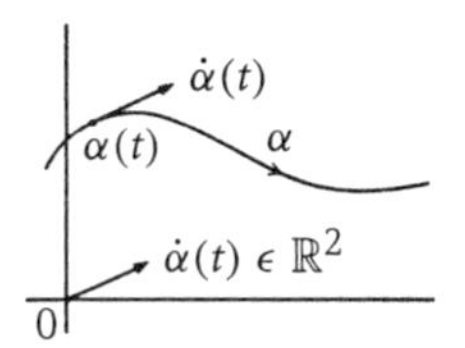

Figure 2.4.

The special case $M \subset \mathbb{R}^N$ will serve as our model for the general case. Of course, it does give the impression at first that the ambient $\mathbb{R}^N$ is what really makes the construction of the tangent space possible! Where else could the tangent spaces live? Nor is extending the definition to arbitrary, "abstract," manifolds a trivial exercise. It involves a certain grandiose way of creating new mathematical objects for which earlier mathematics was, so to speak, too timid.

2.2 Three Versions of the Concept of a Tangent Space

There are three apparently quite different but essentially equivalent definitions of the concept of a *tangent vector*,

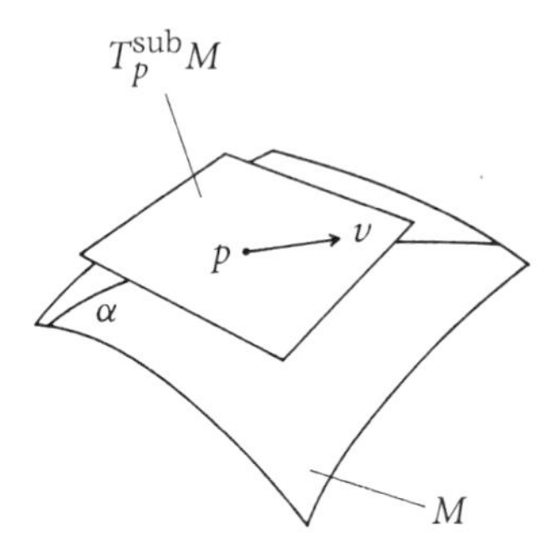

Figure 2.5. For every tangent vector $v \in T_p^{\text{sub}}M$ to a submanifold $M \subset \mathbb{R}^N$, we can find a curve α in M with $\alpha(0) = p$ and $\dot{\alpha}(0) = v$.

which I call (a) the *geometric*, (b) the *algebraic*, and (c) the *"physical"* definitions. We use all three. The order is irrelevant, and we begin with (a).

We start with the intuitive idea of a tangent vector v to a submanifold of $\mathbb{R}^N$ and ask ourselves how we can characterize it without using the ambient space, so as to get a generalizable version of the definition. Well, isn't any such v the velocity vector of a curve α that lies entirely in M, and doesn't such an α contain enough information about v? Actually, too much. Which curves α, β describe the same v? How can we express the equality $\dot{\alpha}(0) = \dot{\beta}(0)$ without using the ambient space $\mathbb{R}^N$? Through charts, for instance: $\dot{\alpha}(0) = \dot{\beta}(0) \in \mathbb{R}^N$ is equivalent to $(h \circ \alpha)^{\cdot}(0) = (h \circ \beta)^{\cdot}(0) \in \mathbb{R}^n$, where (U, h) is a chart on M (!) around p. This is enough motivation for the following definition:

Definition (a). Let M be an n-dimensional manifold, $p \in M$. Let $\mathcal{K}_p(M)$ denote the set of differentiable curves in M that pass through p at $t = 0$; more precisely,

$$\mathcal{K}_p(M) = \{\alpha : (-\varepsilon, \varepsilon) \xrightarrow{C^\infty} M : \varepsilon > 0 \text{ and } \alpha(0) = p\}.$$

Two such curves α, $\beta \in \mathcal{K}_p(M)$ will be called ***tangentially equivalent***, $\alpha \sim \beta$, if

$$(h \circ \alpha)^{\cdot}(0) = (h \circ \beta)^{\cdot}(0) \in \mathbb{R}^n$$

for some (hence any) chart (U, h) around p. We call the equivalence classes $[\alpha] \in \mathcal{K}_p(M)/\sim$ the ***(geometrically defined) tangent vectors*** of M at p, and

$$T_p^{\text{geom}}M := \mathcal{K}_p(M)/\sim$$

the ***(geometrically defined) tangent space*** to M at p.

In preparation for version (b) of the definition, we first introduce the following terminology:

Definition. Let two real-valued functions, each defined and differentiable in some neighborhood of a point p of M, be called equivalent if they agree in a neighborhood of p. The equivalence classes are called the ***germs of differentiable functions on M at p***, and the set of these germs is denoted by $\mathcal{E}_p(M)$.

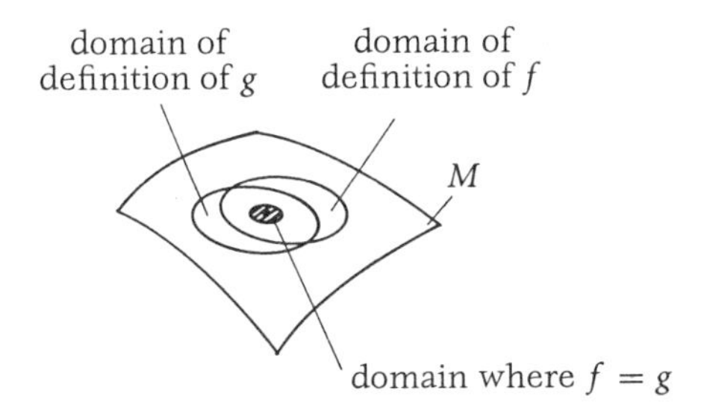

Figure 2.6. For $f \sim g$, f and g need not agree throughout their common domain of definition: a small neighborhood of p suffices.

For convenience, we do not distinguish in our notation between a function $f : U \to \mathbb{R}$ at p and the germ $f \in \mathcal{E}_p(M)$ it represents, and hope no misunderstandings arise. A function f defined at p does contain more information than its germ $[f]$ at p, but the germ is good enough for all those operations where we have to know a function only in *some* neighborhood of p, however small.

It is clear that germs at p can be added and multiplied. More precisely:

Note. *The set $\mathcal{E}_p(M)$ of germs of differentiable functions on M at p is canonically not only a real vector space but also a ring compatible with this vector space structure, and thus a real algebra.*

Now, the starting point for what we call the "algebraic" version of the concept of tangent vectors is the fact that at a given point $p \in \mathbb{R}^n$, a vector $v \in \mathbb{R}^N$ can also be characterized by its directional derivative operator ∇_v at p. To determine $\nabla_v f$ for $v \in T_p^{\text{sub}}M$, all we have to know about f near p is its values *on the submanifold* M, since $\nabla_v f := (f \circ \alpha)^{\cdot}(0)$ for *every* curve α with $\alpha(0) = p$ and $\dot{\alpha}(0) = v$, and we can choose α to lie in M. This leads us to a characterization of v that is independent of the ambient space $\mathbb{R}^N$ and can therefore be generalized.

Definition (b). Let M be an n-dimensional manifold, $p \in M$. By an ***(algebraically defined) tangent vector*** to M at p, we mean a ***derivation*** of the ring $\mathcal{E}_p(M)$ of germs, that is, a linear map

$$v : \mathcal{E}_p \longrightarrow \mathbb{R}$$

that satisfies the product rule

$$v(f \cdot g) = v(f) \cdot g(p) + f(p) \cdot v(g)$$

for all $f, g \in \mathcal{E}_p(M)$. We call the vector space of these derivations the ***(algebraically defined) tangent space*** to M at p and denote it by $T_p^{\text{alg}}(M)$.

Now for version (c). In the physics literature, calculations are generally carried out in coordinates, and usually in a calculus (called the ***Ricci calculus*** in differential geometry) in which the position of the indices (superscripts or subscripts) is significant. What we call a tangent vector is called a ***contravariant vector*** in the Ricci calculus. Briefly, this is an n-tuple, denoted by $(v^1, \ldots, v^n)$ or occasionally (v^0, v^1, v^2, v^3), and abbreviated v^μ, that "transforms" (as we are told) according to the rule

$$\widetilde{v}^\mu = \frac{\partial \widetilde{x}^\mu}{\partial x^\nu} v^\nu.$$

Here, as always in the Ricci calculus, we follow the "summation convention" and, within a term, sum over any index that appears as both superscript and subscript. So we sum here over ν.

What does all this mean? In our language, the following:

Definition (c). Let M be an n-dimensional manifold, $p \in M$. Let $\mathcal{D}_p(M) := \{(U, h) \in \mathcal{D} : p \in U\}$ denote the set of charts around p. By a ***("physically" defined) tangent vector*** v to M at p, we mean a map

$$v : \mathcal{D}_p(M) \to \mathbb{R}^n$$

with the property that for any two charts the associated vectors in $\mathbb{R}^n$ are mapped to each other by the differential of the transition map; that is,

$$v(V,k) = d(k \circ h^{-1})_{h(p)} \cdot v(U,h)$$

for all (U,h), $(V,k) \in \mathcal{D}_p(M)$. We call the vector space of these maps v the ***(physically defined) tangent space*** to M at p and denote it by $T_p^{\text{phys}}M$.

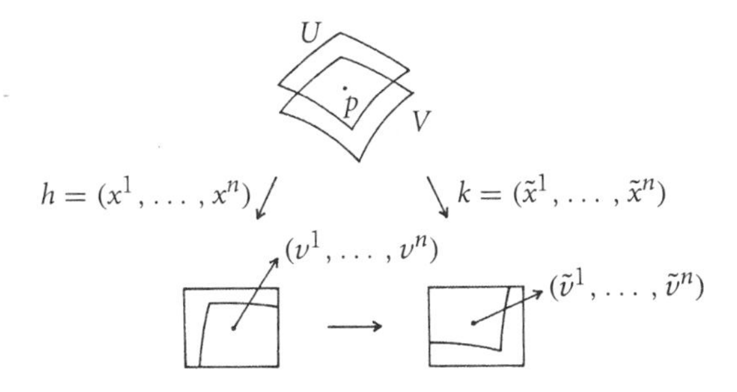

Figure 2.7. Interpreting the transformation law $\widetilde{v}^\mu = \frac{\partial \widetilde{x}^\mu}{\partial x^\nu} v^\nu$ for "contravariant vectors": $\frac{\partial \widetilde{x}^\mu}{\partial x^\nu}\big|_{h(p)}$ is the Jacobian matrix of the transition map $\widetilde{x}^\mu = \widetilde{x}^\mu(x^1, \dots, x^n)$, $\mu = 1, \dots, n$.

By the way, I don't mean to make fun of the Ricci calculus. It's a very elegant calculus that guides the user through explicit computations—virtually a *machine-readable* calculus—and is constantly used in the physics literature because there is still no better practical calculus for vector and tensor analysis. But these advantages, which you'll learn to value more highly on closer acquaintance, come at the cost of some disadvantages. The elegance of a system of notation is usually based on the suppression of "unimportant" data, and different things are important for the efficient manipulation of formulas than for the logical explanation of fundamental geometric concepts. So, for now, we have to denote a "contravariant vector" not by a graceful and economical v^μ but with the unwieldy precision of

$$v : \mathcal{D}_p(M) \to \mathbb{R}^n, \quad (U,h) \mapsto v(U,h).$$

This is not intended as a recommendation for improvement of the physicists' everyday practice.

2.3 Equivalence of the Three Versions

We now want to convince ourselves that the three versions of the concept of a tangent space are essentially the same. The following lemma should not be seen as punishment for our having willfully defined the same thing three times over, but as a whole system of indispensable lemmas about the tangent space, which are most clearly summarized in this form.

Lemma. *The canonical maps*

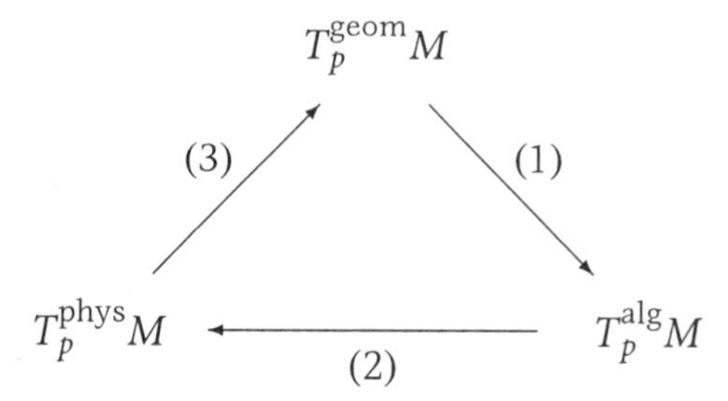

described more fully below are mutually compatible bijections; that is, the composition of any two is the inverse of the third.

PRECISE STATEMENT AND PROOF. We first give the three maps.

(1) Geometric $\longrightarrow$ algebraic. *If $[\alpha]$ is a geometrically defined tangent vector to M at p, then the map*

$$\begin{aligned} \mathcal{E}_p(M) &\longrightarrow \mathbb{R}, \\ f &\longmapsto (f \circ \alpha)^{\cdot}(0) \end{aligned}$$

is a derivation and therefore an algebraically defined tangent vector. Of course, this takes a few little proofs. That the map is independent of the choice of representative function for the germ is clear and has been legitimately anticipated in our notation. To check that $(f \circ \alpha)^{\cdot}(0)$ is independent of the choice of representative $\alpha \in \mathcal{K}_p(M)$, we use a chart (U, h) around p. Without loss of generality, we may assume that

$f : U \to \mathbb{R}$ represents the germ and that α and β have the same sufficiently small domain of definition $(-\varepsilon, \varepsilon)$. Then $(h \circ \alpha)^{\cdot}(0) = (h \circ \beta)^{\cdot}(0)$ by hypothesis; hence

$$(f \circ h^{-1} \circ h \circ \alpha)^{\cdot}(0) = (f \circ h^{-1} \circ h \circ \beta)^{\cdot}(0)$$

by the chain rule.

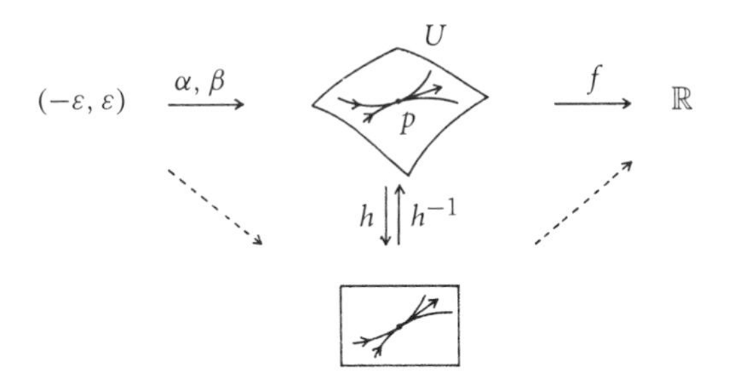

Figure 2.8. By the chain rule, tangentially equivalent curves define the same derivation $f \mapsto (f \circ \alpha)^{\cdot}(0)$.

Finally, the product rule for functions $(-\varepsilon, \varepsilon) \to \mathbb{R}$ implies that the map $\mathcal{E}_p(M) \to \mathbb{R}$, $f \mapsto (f \circ \alpha)^{\cdot}(0)$, which we now know is well defined for a given $[\alpha]$, is really a derivation.

(2) Algebraic $\longrightarrow$ physical. *If $v : \mathcal{E}_p(M) \to \mathbb{R}$ is a derivation, then the map*

$$\begin{aligned} \mathcal{D}_p(M) &\longrightarrow \mathbb{R}^n, \\ (U, h) &\longmapsto (v(h_1), \dots, v(h_n)) \end{aligned}$$

is a physically defined tangent vector. If (U, h) and (V, k) are charts around p and $w := k \circ h^{-1}$ on $h(U \cap V)$ is the transition map, then we have to show that

$$(*) \qquad v(k_i) = \sum_{j=1}^{n} \frac{\partial w_i}{\partial x_j}(h(p)) \cdot v(h_j).$$

Now, this is the only place in our study of the relationships among the three definitions of the tangent space where we really need a little trick.

All we know about v is that it is a derivation. We should therefore try to arrive somehow at a representation of the form

$$k_i = \sum_{j=1}^{n} g_{ij} \cdot h_j,$$

so that we can also take advantage of the product rule. This works, with the following lemma.

AUXILIARY LEMMA. *Let $\Omega \subset \mathbb{R}^n$ be an open set that is star-shaped with respect to the origin (an open ball around 0, for instance, or $\mathbb{R}^n$ itself). If $f : \Omega \to \mathbb{R}$ is a differentiable (i.e. C^∞) function with $f(0) = 0$, then there exist differentiable functions $f_j : \Omega \to \mathbb{R}$ with*

$$f(x) = \sum_{j=1}^{n} x_j \cdot f_j(x).$$

PROOF OF THE AUXILIARY LEMMA. Since

$$f(x) = \int_0^1 \frac{d}{dt} f(tx_1, \ldots, tx_n)dt = \int_0^1 \sum_{j=1}^{n} x_j \frac{\partial f}{\partial x_j}(tx_1, \ldots, tx_n)dt,$$

we need only set

$$f_j(x) := \int_0^1 \frac{\partial f}{\partial x_j}(tx_1, \ldots, tx_n)dt. \qquad \square$$

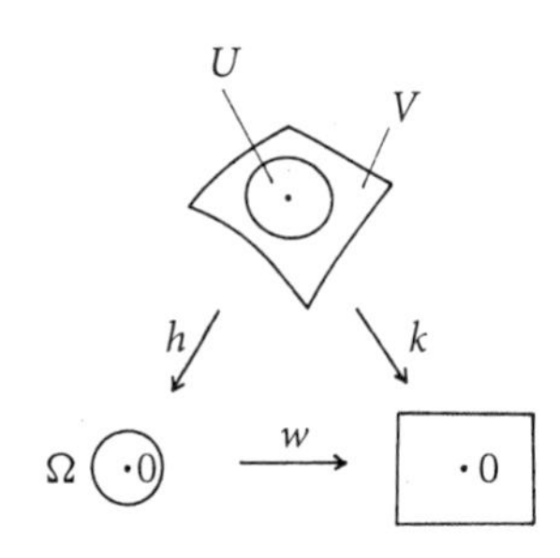

Figure 2.9. Transition map on an open ball Ω around 0

APPLICATION OF THE AUXILIARY LEMMA. Without loss of generality, we may assume that $h(p) = k(p) = 0$ and $h(U)$ is an open ball around 0 so small that U is contained in V. Then according to our lemma the n component functions $w_1, \ldots, w_n$ of the transition map have the form

$$w_i = \sum_{j=1}^{n} x_j w_{ij}(x),$$

and since $k = w \circ h$ it follows that

$$k_i = \sum_{j=1}^{n} (w_{ij} \circ h) \cdot h_j,$$

as we hoped. If we now apply the derivation v to k, then

$$v(k_i) = \sum_{j=1}^{n} (w_{ij} \circ h)(p) \cdot v(h_j) = \sum_{j=1}^{n} w_{ij}(0) \cdot v(h_j)$$

because $h(p) = 0$. But $w_{ij}(0)$ is just $\frac{\partial w_i}{\partial x_j}(0)$, and we have verified (*).

(3) Physical $\longrightarrow$ geometric. *If $v : \mathcal{D}_p \to \mathbb{R}^n$ is a physically defined tangent vector and (U, h) a chart around p, and if $\alpha : (-\varepsilon, \varepsilon) \to U$ is defined for sufficiently small $\varepsilon > 0$ by*

$$\alpha(t) := h^{-1}(h(p) + tv(U, h)),$$

then $[\alpha] \in T_p^{\text{geom}} M$ is independent of the choice of chart. More precisely, let β be the corresponding curve in terms of (V, k) and w the transition map. If we use k to test the tangential equivalence of α and β, then $(k \circ \alpha)^{\boldsymbol{\cdot}}(0) = (k \circ \beta)^{\boldsymbol{\cdot}}(0)$ is equivalent to $dw_{h(p)}(v(U, h)) = v(V, k)$. But this is just the transformation law for the physically defined tangent vector v, by definition.

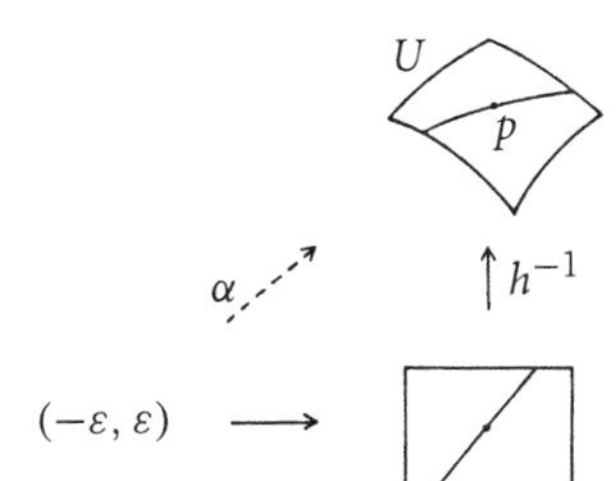

Figure 2.10. Defining the map $T_p^{\text{phys}} M \to T_p^{\text{geom}} M$

We have now explicitly given the three maps described as canonical in the lemma, and we denote them by Φ_1, Φ_2, and Φ_3.

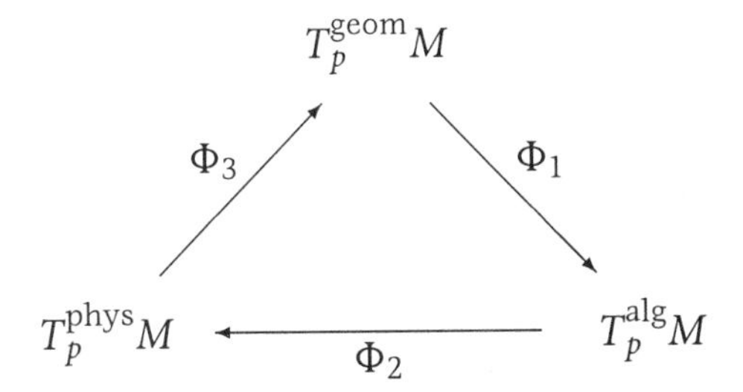

It remains to show that any circuit around the diagram yields the identity, or more precisely that

$$\begin{aligned}\Phi_3 \circ \Phi_2 \circ \Phi_1 &= \mathrm{Id}_{T_p^{\mathrm{geom}}M},\\ \Phi_2 \circ \Phi_1 \circ \Phi_3 &= \mathrm{Id}_{T_p^{\mathrm{phys}}M},\\ \Phi_1 \circ \Phi_3 \circ \Phi_2 &= \mathrm{Id}_{T_p^{\mathrm{alg}}M}.\end{aligned}$$

A geometric tangent vector $[\alpha]$, for instance, first becomes the derivation $f \mapsto (f \circ \alpha)^{\cdot}(0)$, and this becomes the physical vector $v(U, h) = (h \circ \alpha)^{\cdot}(0)$ we finally use to construct the curve $\beta(t) := h^{-1}(h(p) + t(h \circ \alpha)^{\cdot}(0))$ that represents the geometric tangent vector $\Phi_3 \circ \Phi_2 \circ \Phi_1[\alpha]$. Is it true that $[\beta] = [\alpha]$? Yes, because it is clear that $(h \circ \beta)^{\cdot}(0)$ is just $(h \circ \alpha)^{\cdot}(0)$. The other two formulas can be verified similarly, and with this assurance we end the proof of the lemma. □

2.4 Definition of the Tangent Space

Now that it's clear in what way $T_p^{\mathrm{geom}}M$, $T_p^{\mathrm{alg}}M$, and $T_p^{\mathrm{phys}}M$ are actually the same object, how shall we define the *tangent space in general*? Should I just say, *"Let's call it T_pM"*? A mysterious archetype, of which the three real versions are only fleeting likenesses? Preferably not. Or shall we somehow *identify* the three versions into a single T_pM by taking equivalence classes? A bit better, maybe, but what's the point? Aren't three versions enough? Do we really have to come up with a fourth?

The actual (and sensible) practice is to use all three versions simultaneously and indiscriminately but omit their labels, with the tacit understanding that exactly which version is being used at the time is either obvious or unimportant. But to help you answer the legitimate question "What is a tangent vector?" without being forced into lengthy explanations to yourself or anyone else, we proceed a bit more formally and make the following definition.

Definition. Let M be an n-dimensional manifold, $p \in M$. The vector space

$$T_pM := T_p^{\mathrm{alg}}M$$

will be called the ***tangent space to M at p***, and its elements ***tangent vectors***. We agree, though, that whenever necessary we will also consider a derivation $v \in T_pM$ as a geometrically or physically defined tangent vector, as in Section 2.3, and denote it by the same symbol if there is no risk of confusion.

Note. *If M is an n-dimensional manifold, then the canonical bijection $T_p^{\mathrm{alg}}M \cong T_p^{\mathrm{phys}}M$ is linear, and for a fixed chart, the map $v \mapsto v(U, h)$ defines an isomorphism $T_p^{\mathrm{phys}}M \cong \mathbb{R}^n$. Hence the tangent spaces to M are also n-dimensional.*

2.5 The Differential

I described the introduction of the tangent space as a preliminary to defining the *differential*, the local linear approximation of a differentiable map between manifolds. Now the preliminary work is finished and we turn to the differential. Although I have no intention of presenting all future definitions regarding tangent spaces in triplicate, it should be done one more time. Let $f : M \to N$ be a differentiable map, $p \in M$. Let's consider in turn the geometric, algebraic, and physical versions of how f canonically induces a linear map between the tangent spaces at p and $f(p)$.

On geometric tangent vectors, f acts by *curve transport:*

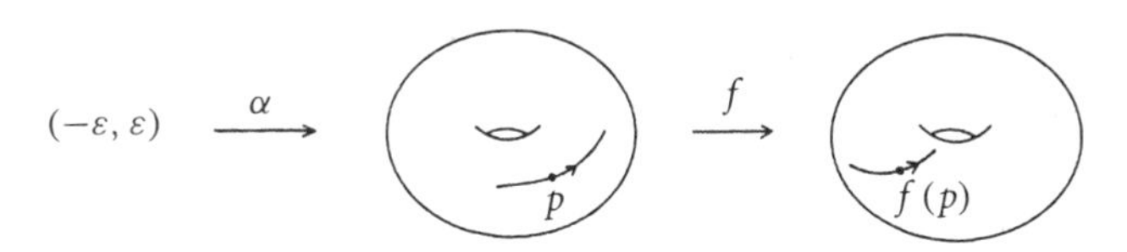

Figure 2.11. The curve $\alpha \in \mathcal{K}_p(M)$ is "transported" by f to the curve $f \circ \alpha \in \mathcal{K}_{f(p)}(N)$.

You can easily check that the map

$$\begin{aligned} d^{\mathrm{geom}} f_p : T_p^{\mathrm{geom}} M &\longrightarrow T_{f(p)}^{\mathrm{geom}} N, \\ [\alpha] &\longmapsto [f \circ \alpha] \end{aligned}$$

is well defined.

What about algebraic tangent vectors? Precomposing by f assigns germs at p to germs at $f(p)$

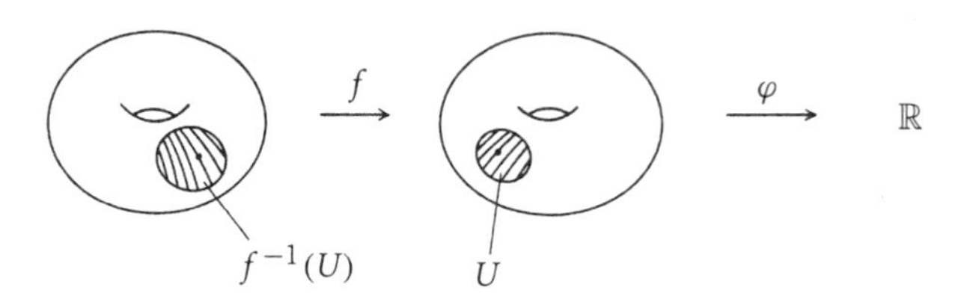

Figure 2.12. The germ of $\varphi \circ f | f^{-1}(U)$ at p is assigned to the germ of $\varphi : U \to \mathbb{R}$ at $f(p)$.

and thus defines an *algebra homomorphism*

$$\begin{aligned} f^* : \mathcal{E}_{f(p)}(N) &\longrightarrow \mathcal{E}_p(M), \\ \varphi &\longmapsto \varphi \circ f. \end{aligned}$$

Precomposing by f^* turns a derivation at p into a derivation at $f(p)$: the map

$$\begin{aligned} d^{\text{alg}} f_p : T_p^{\text{alg}} M &\longrightarrow T_{f(p)}^{\text{alg}} N \\ v &\longmapsto v \circ f^* \end{aligned}$$

is well defined and obviously linear. Therefore $d^{\text{alg}} f_p$ acts as a derivation on germs φ at $f(p)$ by $\varphi \mapsto v(\varphi \circ f)$.

Finally, to describe the linear map

$$d^{\text{phys}} f_p : T_p^{\text{phys}} M \longrightarrow T_{f(p)}^{\text{phys}} N$$

canonically induced by f between the "physically" defined tangent spaces, we must exhibit each

$$(d^{\text{phys}} f_p(v))(V, k) \in \mathbb{R}^{\dim N}.$$

To do this we choose a chart (U, h) around p with $f(U) \subset V$ and set

$$(d^{\text{phys}} f_p(v))(V, k) := d(k \circ f \circ h^{-1})_{h(p)} v(U, h),$$

which just means that $d^{\text{phys}} f_p$ is given, in terms of charts, by the Jacobian matrix of the downstairs map. You can use the chain rule to check that this is well defined.

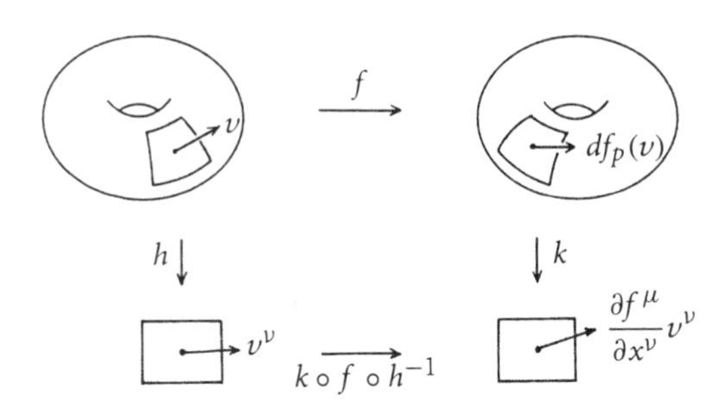

Figure 2.13. The differential in the Ricci calculus: the contravariant vector v^ν goes to $\frac{\partial f^\mu}{\partial x^\nu} \cdot v^\nu$.

Lemma and Definition. *Let $f : M \to N$ be a differentiable map between manifolds, $p \in M$. The three maps*

$$\begin{aligned} d^{\text{geom}} f_p : T_p^{\text{geom}} M &\longrightarrow T_{f(p)}^{\text{geom}} N, \\ d^{\text{alg}} f_p : T_p^{\text{alg}} M &\longrightarrow T_{f(p)}^{\text{alg}} N, \\ d^{\text{phys}} f_p : T_p^{\text{phys}} M &\longrightarrow T_{f(p)}^{\text{phys}} N \end{aligned}$$

induced by f through curve transport, homomorphism of algebras of germs, and (in terms of charts) the Jacobian matrix are compatible with the canonical bijections between the geometric, algebraic, and physical tangent spaces, respectively. Hence they all define the same linear map

$$df_p : T_p M \longrightarrow T_{f(p)} N,$$

*which we call the **differential of f at the point p.***

Since the proof consists of arguments that are familiar by now, I won't write it out. But this doesn't mean that the statement is *obvious*. It takes some experience to believe the lemma from conviction rather than just on authority, and if we were the first ones to be interested in it we would have to look it over pretty carefully to see if the devil's hiding somewhere in the details. What is clear in all three versions is the functorial property of the differential, which we record as the most important property so far of the newly defined concept.

Note. *The differential of the identity is the identity,*

$$d\,\mathrm{Id}_p = \mathrm{Id}_{T_pM},$$

and the ***chain rule*** *holds; that is,*

$$d(g \circ f)_p = dg_{f(p)} \circ df_p$$

for the composition $M_1 \xrightarrow{f} M_2 \xrightarrow{g} M_3$ *of differentiable maps.*

This temporarily concludes our introduction to the basic concepts of differential topology; we consider differential forms in the next chapter. But in the next three sections of the present chapter, we still have to discuss a few questions of notation.

2.6 The Tangent Spaces to a Vector Space

Every n-dimensional real vector space V is canonically an n-dimensional manifold; its topology and differentiable structure are characterized by the requirement that the isomorphisms $V \cong \mathbb{R}^n$ must also be *diffeo*morphisms. Of course, the motive we gave for introducing tangent spaces is not valid in this special case: a linear space doesn't have to be approximated linearly. So we can hardly be surprised that for every $p \in V$ there is a *canonical* isomorphism

$$V \xrightarrow{\cong} T_pV.$$

The tangent vector this assigns to a vector $v \in V$ is given geometrically, for example, by the curve

$$t \mapsto p + tv,$$

and hence algebraically by the derivation

$$\varphi \mapsto \left.\frac{d}{dt}\right|_0 \varphi(p + tv).$$

If we consider the elements $v \in V$ as tangent vectors in this way, then the differential at p of a differentiable map $f : V \to W$ between finite-dimensional real vector spaces becomes a linear map

$$df_p : V \longrightarrow W,$$

and this is how we usually write it, in particular for $V = \mathbb{R}^n$, $W = \mathbb{R}^k$. The notation $T_p\mathbb{R}^n$ will be used only when clarity demands it. So the differential, viewed as a linear map $df_p : \mathbb{R}^n \to \mathbb{R}^k$, is just given by the Jacobian matrix $J_f(p)$.

Of course, we often refer implicitly to $T_p\mathbb{R}^n$ and T_pV, since the special cases $M = \mathbb{R}^n$ and $M = V$ are also present in any discussion of T_pM for arbitrary M. We have no intention of dismissing the tangent spaces of a vector space.

2.7 Velocity Vectors of Curves

For each value of the parameter $t \in (a, b)$, a differentiable curve $\alpha : (a, b) \to M$ has a ***velocity vector***, which we denote by $\dot{\alpha}(t) \in T_{\alpha(t)}M$. To be precise, $\dot{\alpha}(t)$ is represented geometrically by $\lambda \mapsto \alpha(t + \lambda)$, algebraically by the derivation $\varphi \mapsto (\varphi \circ \alpha)^{\cdot}(t)$, and as a physical tangent vector (in local coordinates) by $(U, h) \mapsto (h \circ \alpha)^{\cdot}(t)$. The notation $\dot{\alpha}(t)$ actually comes from curves in $\mathbb{R}^n$, where of course it stands for

$$\dot{\alpha}(t) = (\dot{\alpha}_1(t), \dots, \dot{\alpha}_n(t)) \in \mathbb{R}^n.$$

But no notational clashes occur because the isomorphism $\mathbb{R}^n \cong T_{\alpha(t)}\mathbb{R}^n$ takes this ordinary $\dot{\alpha}(t) \in \mathbb{R}^n$ to our newly defined $\dot{\alpha}(t) \in T_{\alpha(t)}\mathbb{R}^n$.

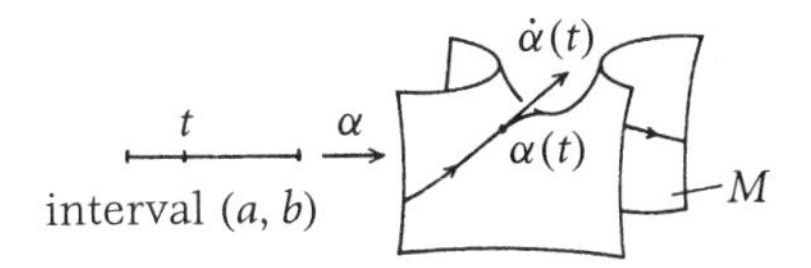

Figure 2.14. The concept of the velocity vector $\dot{\alpha}(t) \in T_{\alpha(t)}M$

Observe that instead of $[\alpha] \in T_p^{\text{geom}}M$ we may (and will) write $\dot{\alpha}(0)$, and that the description of the differential by means of curve transport gives

$$df_{\alpha(t)}(\dot{\alpha}(t)) = (f \circ \alpha)^{\cdot}(t),$$

where α denotes a curve in M and $f : M \to N$ a differentiable map.

2.8 Another Look at the Ricci Calculus

A chart (U, h) introduces ***coordinates*** on the chart domain U. These are just the component functions $h_1, \ldots, h_n$ of the chart map $h = (h_1, \ldots, h_n)$. Each individual coordinate is thus a real function $h_i : U \to \mathbb{R}$, and a point $p \in U$ has coordinates $(h_1(p), \ldots, h_n(p))$. For each $p \in U$ the chart also distinguishes a basis of T_pM, namely the one that corresponds to the canonical basis $(e_1, \ldots, e_n)$ of $\mathbb{R}^n$ under the map

$$\begin{aligned} T_p^{\text{phys}}M &\xrightarrow{\cong} \mathbb{R}^n, \\ v &\longmapsto v(U, h). \end{aligned}$$

I would like to introduce notation for this basis and, in doing so, pick up where I left off in Section 2.2 and say a bit more about the Ricci calculus.

Its unmatched elegance for computations with geometric objects in local coordinates has already been praised. With a minimum of arbitrary notation (admittedly with a lot of indices), it describes all the local objects and procedures of vector and tensor analysis in such a way that one can insert numbers at any time and start computing. Moreover, the notation always automatically indicates behavior under transformations—for the expert, the geometric nature of things. The calculus *thinks for the user.* But if we want to enjoy these benefits, we have to put up with some unpleasantness. Let's start with the rather harmless rituals for entering this temple.

We set down the notation U for the chart domain at the entrance. It's obvious, says the Ricci calculus, that a coordinate system has a certain domain of validity, so let's not waste letters. Next we are asked to *raise* the indices of the coordinates, to write them as

$$h = (h^1, \dots, h^n).$$

We do it none too willingly because superscripts usually denote exponents, but this goes best with the index conventions of the calculus, where superscripts can't be avoided anyway. So we do it. But now the letter h is also discarded as arbitrary and uninformative. The coordinates will be written

$$x^1, \dots, x^n,$$

so they can be immediately recognized as coordinates. If we ever have to deal with another coordinate system, we can mark its coordinates somehow to distinguish them, say by writing them as

$$\widetilde{x}^1, \dots, \widetilde{x}^n,$$

and if a second manifold enters the picture, then coordinates

$$y^1, \dots, y^k,$$

etc., will also be allowed. But the first choice of names for coordinates is still $x^1, \dots, x^n$. In this interpretation—if we also secretly use the forbidden letters U and h to make things clear—the coordinates become functions $x^\mu : U \to \mathbb{R}$, so that $h = (x^1, \dots, x^n)$ holds. This clash is not unintended by the calculus, just as when, in older textbooks on differential calculus, a real function is written as

$$y = y(x),$$

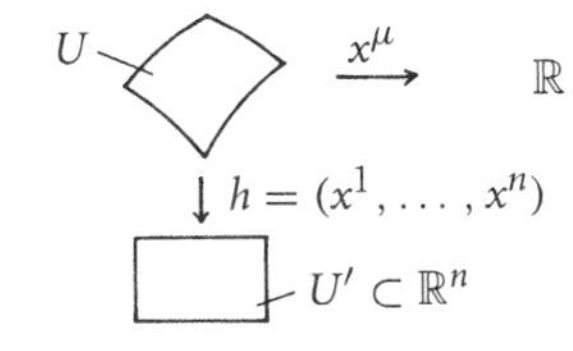

Figure 2.15. The coordinates x^μ in the Ricci calculus

which has the advantage (and disadvantage) that then the function doesn't need (or have) a name of its own. But in any case this *is* a double meaning of x^μ as both a function on $U \subset M$ and a coordinate of $\mathbb{R}^n$, and we have to keep it in mind, especially because we now fix the following notation.

Notation. Let $p \in U$, where (U, h) is a chart with coordinates $x^1, \dots, x^n$, i.e. $h = (x^1, \dots, x^n)$. Then the μth vector of the basis of T_pM given by the coordinates will be denoted by

$$\frac{\partial}{\partial x^\mu} \in T_pM$$

and abbreviated $\partial_\mu \in T_pM$.

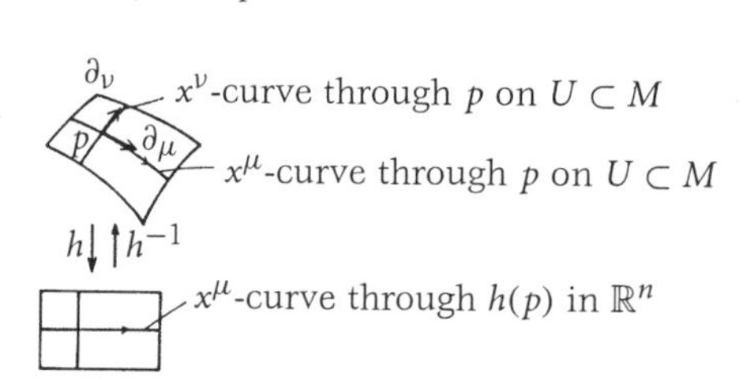

Figure 2.16. Coordinate basis $(\partial_1, \dots, \partial_n)$ of T_pM

To prevent misunderstandings: ∂_μ, as a physical tangent vector in $T_p^{\text{phys}}M$, just assigns the μth unit vector $e_\mu \in \mathbb{R}^n$ to our chart (U, h); as a geometric tangent vector $\partial_\mu \in T_p^{\text{geom}}M$, it is represented by means of the curve $t \mapsto h^{-1}(h(p) + te_\mu)$ (∂_μ is the velocity vector of the μth coordinate curve); and finally, as a derivation, ∂_μ acts by

$$\begin{aligned} \mathcal{E}_p(M) &\longrightarrow \mathbb{R}, \\ \varphi &\longmapsto \frac{\partial(\varphi \circ h^{-1})}{\partial x^\mu}(h(p)), \end{aligned}$$

and thus as the μth partial derivative of the downstairs function. And this is just what the Ricci notation $\partial_\mu\varphi$, despite its superb terseness, suggests unambiguously. What possible meaning could applying $\partial/\partial x^\mu$ to a function φ defined on a manifold have, other than *first expressing the function in the coordinates* $x^1, \dots, x^n$, i.e. taking $\varphi \circ h^{-1}$, then differentiating with respect to the μth coordinate?

You may object that the notation contains no information about p. How can you tell $\partial_\mu \in T_pM$ from $\partial_\mu \in T_qM$? Well, if we wanted to indicate which tangent space we were in at the moment, we would have to resort to an additional label, say $\partial_\mu|_p$. But this is hardly ever necessary. We often have no fixed p in mind at all, but rather the assignment to *every* $p \in U$ of its $\partial_\mu \in T_pM$, and for this *vector field* on U the notation ∂_μ or

Figure 2.17. The vector field ∂_μ on U

$\partial/\partial x^\mu$ fits perfectly. The tangent vector

$$v^\mu \partial_\mu := v^1 \partial_1 + \cdots + v^n \partial_n$$

corresponds to the "contravariant vector" written in the Ricci calculus as v^μ, and the context will have to tell whether a fixed $p \in U$ is being considered and $v^\mu \partial_\mu \in T_p M$ is meant, or, as occurs more often, the $v^1, \ldots, v^n$ are real functions on U, and $v^\mu \partial_\mu$ denotes a vector *field* on U.

2.9 Test

(1) For the two "poles" $p := (0, 0, 1)$ and $q := (0, 0, -1)$ of the 2-sphere $S^2 \subset \mathbb{R}^3$, it is obvious that

$$T_p^{\text{sub}} S^2 = T_q^{\text{sub}} S^2 = \mathbb{R}^2 \times \{0\} \subset \mathbb{R}^3.$$

Is it also true that $T_p^{\text{geom}} S^2 = T_q^{\text{geom}} S^2$, and similarly for "alg" and "phys"?

- ☐ Yes, because canonically $T_p^{\text{sub}} M \cong T_p^{\text{geom}}$, etc.
- ☐ No. For all three versions, $T_p M \cap T_q M = \emptyset$ if $p \neq q$.
- ☐ Yes for T^{phys}, no for the other two versions, because in those cases $T_p M \cap T_q M = \{0\}$ if $p \neq q$.

(2) Do two functions f and g defined at 0 in $\mathbb{R}^n$ define the same germ in $\mathcal{E}_0(\mathbb{R}^n)$ if they have the same partial derivatives of every degree at 0?

- ☐ No. (Hint: e^{-1/x^2}.)
- ☐ Yes, by Taylor's formula for functions of several variables.
- ☐ Yes. Otherwise we would get a contradiction to the mean value theorem.

(3) Let $M_0 \subset M$ be a submanifold, $p \in M_0$, and let $v \in T_p M$ be a derivation such that $vf = 0$ for all $f \in \mathcal{E}_p(M)$ that vanish on M_0. Then

☐ $v \in T_pM_0 \subset T_pM$.

☐ $v = 0$.

☐ $v \in T_pM \setminus T_pM_0$.

(4) For differentiable maps f between manifolds, we have

☐ $\operatorname{rk} df_p = \operatorname{rk}_p f$, always.

☐ $\operatorname{rk} df_p \geq \operatorname{rk}_p f$, and $>$ can occur.

☐ $\operatorname{rk} df_p \leq \operatorname{rk}_p f$, and $<$ can occur.

(5) Let $f : M \to N$ be constant. Then $df_p =$

☐ $f(p)$. ☐ 0. ☐ Id_{T_pM}.

(6) Let $f : V \to W$ be a linear map between finite-dimensional real vector spaces V and W. Then $df_p =$

☐ f. ☐ 0. ☐ $f - f(p)$.

(7) Let V be a finite-dimensional real vector space and let $f : V \to V$ be a translation. Then $df_p =$

☐ f. ☐ 0. ☐ Id_V.

(8) Let M be a differentiable manifold and let X, Y, and Z be finite-dimensional vector spaces. Also let $\langle \cdot, \cdot \rangle : X \times Y \to Z$ denote some bilinear operation. Then for differentiable maps $f : M \to X$ and $g : M \to Y$ we have, at every point $p \in M$,

☐ $d\langle f, g\rangle = \langle df, g\rangle + \langle f, dg\rangle$.

☐ $d\langle f, g\rangle = \langle df, g\rangle - \langle f, dg\rangle$.

☐ $d\langle f, g\rangle = \langle df, dg\rangle$.

(9) Let a differentiable map $f : M \to N$ be described in local coordinates $x^{\overline{\nu}}$ for N and x^{μ} for M by

$$x^{\overline{\nu}} = x^{\overline{\nu}}(x^1, \ldots, x^n),$$

in the sense of the Ricci calculus. Then the matrix of the differential is given by

☐ $\partial_\mu x^\nu$. ☐ $\partial_\mu x^{\overline{\nu}}$. ☐ $\partial_{\overline{\nu}} x^\mu$.

(10) Under what additional hypotheses on a map $f : M \to N$ can arbitrary vector fields be transported canonically from one manifold to another by the differentials df_p or their inverses?

☐ It's always possible from M to N, but it's possible in the opposite direction only if f is a covering map.

☐ Even from M to N, it's possible only if f is a diffeomorphism.

☐ It's possible in both directions as long as f is an embedding.

2.10 Exercises

EXERCISE 2.1. Let M be an n-dimensional manifold, $p \in M$. Show that the composition of the canonical maps

$$T_p^{\mathrm{alg}} M \to T_p^{\mathrm{phys}} M \to T_p^{\mathrm{geom}} M \to T_p^{\mathrm{alg}} M$$

is the identity on $T_p^{\mathrm{alg}} M$.

EXERCISE 2.2. Let $f : M \to N$ be a differentiable map, $p \in M$. Prove that the diagram

$$\begin{array}{ccc} T_p^{\mathrm{geom}} M & \xrightarrow{d^{\mathrm{geom}} f_p} & T_{f(p)}^{\mathrm{geom}} N \\ \downarrow & & \downarrow \\ T_p^{\mathrm{alg}} M & \xrightarrow{d^{\mathrm{alg}} f_p} & T_{f(p)}^{\mathrm{alg}} N \end{array}$$

is commutative.

EXERCISE 2.3. Let $f : M \to \mathbb{R}$ be a differentiable function, $p \in M$. Taking the gradient relative to charts gives a map

$$\begin{array}{rcl} \mathcal{D}_p(M) & \longrightarrow & \mathbb{R}^n, \\ (U, h) & \longmapsto & \mathrm{grad}_{h(p)}(f \circ h^{-1}). \end{array}$$

Call it $\text{grad}_p f$. Is it an element of $T_p^{\text{phys}}M$?

EXERCISE 2.4. Let $M_0 \subset M$ be a submanifold, $p \in M_0$. Canonically, namely through the differential of the inclusion $M_0 \hookrightarrow M$, we regard T_pM_0 as a subspace of T_pM. Show that if M_0 is the preimage of a regular value of a map $f : M \to N$, then

$$T_pM_0 = \ker df_p.$$

2.11 Hints for the Exercises

FOR EXERCISE 2.1. Although the three maps are canonical and therefore independent of charts, a chart (U, h) does come up as a tool in describing $T_p^{\text{phys}}M \to T_p^{\text{geom}}M$. So the proof should start like this: Let (U, h) be a chart around p and let $v \in T_p^{\text{alg}}M$ be a derivation. Then the derivation $v' := \Phi_1(\Phi_3(\Phi_2(v)))$ is given by $v'\varphi = \dots$—and the first part of the exercise will just consist in computing this from the definitions of the Φ_i, which you know.

For the second part, the proof that $v'\varphi = v\varphi$, you might start by justifying why you may assume without loss of generality that $\varphi(p) = 0$ and $h(p) = 0$ (note that $v(\text{const}) = 0$ for any derivation). Then apply the auxiliary lemma from the proof in Section 2.3.

FOR EXERCISE 2.2. All you have to do is compare the fates of a geometric tangent vector ("Let $[\alpha] \in T_p^{\text{geom}}M \dots$") along the two paths to the lower right-hand corner. This is easier than Exercise 2.1.

FOR EXERCISE 2.3. It's certainly true that $\text{grad}_p f$ is sometimes a vector, for instance when $f \equiv 0$. In general? Proof or counterexample? But who can doubt that the gradient is a tangent vector? Or is it?

FOR EXERCISE 2.4. In proving that two vector spaces are equal, if you have some prior knowledge of the dimensions you can often get by with proving just one of the two inclusions.

3 CHAPTER Differential Forms

3.1 Alternating k-Forms

Differential forms live on manifolds, and in preparation for the definition we need some linear algebra in a real vector space that we will later specialize to T_pM.

Definition. Let V be a real vector space. An ***alternating k-form*** ω on V is a map

$$\omega : \underbrace{V \times \cdots \times V}_{k} \longrightarrow \mathbb{R}$$

that is multilinear (i.e. linear in each of the k variables) and has the additional property that $\omega(v_1, \dots, v_k) = 0$ if $v_1, \dots, v_k \in V$ are linearly dependent.

Notation. The vector space of alternating k-forms on V will be denoted by $\mathrm{Alt}^k V$.

Clearly it *is* a real vector space in a canonical way. Strictly speaking, the wording of the definition assumes that $k \geq 1$, but we can extend it appropriately to $k = 0$.

Convention. $\mathrm{Alt}^0 V := \mathbb{R}$.

Thus the alternating 0-forms are the real numbers. $\mathrm{Alt}^1 V = \mathrm{Hom}(V, \mathbb{R}) =: V^*$ is the usual ***dual space*** of V. For $k = 1$ the "alternating" property is trivial since linearity gives it immediately: $\omega(0) = 0$. But for $k \geq 2$, to be alternating means something special, and a few criteria will be useful.

Lemma. *For multilinear maps $\omega : V \times \cdots \times V \to W$, the following conditions are equivalent:*

(1) *ω is alternating; that is, $\omega(v_1, \ldots, v_k) = 0$ if $v_1, \ldots, v_k$ are linearly dependent.*

(2) *$\omega(v_1, \ldots, v_k) = 0$ if any two of the v_i are equal, that is, if there are indices i, j with $i \neq j$ and $v_i = v_j$.*

(3) *Interchanging two of the variables switches the sign:*

$$\omega(v_1, \ldots, v_k) = -\omega(v_1, \ldots, v_j, \ldots, v_i, \ldots, v_k) \text{ for } i < j.$$

(4) *If $\tau : \{1, \ldots, k\} \to \{1, \ldots, k\}$ is a permutation, then*

$$\omega(v_{\tau(1)}, \ldots, v_{\tau(k)}) = \mathrm{sgn}(\tau)\omega(v_1, \ldots, v_k).$$

PROOF. The implications (1) $\Rightarrow$ (2) $\Leftarrow$ (3) $\Leftrightarrow$ (4) are trivial. (2) $\Rightarrow$ (1) is also immediate: if $v_1, \ldots, v_k$ are linearly dependent, then one of the vectors is a linear combination of the others, so $\omega(v_1, \ldots, v_k)$ becomes a sum whose $k - 1$ summands all vanish because of (2). For (2) $\Rightarrow$ (3), observe that (2) implies not only that

$$\omega(v_1, \ldots, v_i + v_j, \ldots, v_i + v_j, \ldots, v_k) = 0$$

but also that, of the four summands linearity gives on the left-hand side, only two remain. This gives

$$\omega(v_1, \ldots, v_i, \ldots, v_j, \ldots, v_k) + \omega(v_1, \ldots, v_j, \ldots, v_i, \ldots, v_k) = 0$$

and hence (3). □

A linear map $f : V \to W$ induces a linear map $\mathrm{Alt}^k f : \mathrm{Alt}^k W \to \mathrm{Alt}^k V$ in the "opposite direction"; thus Alt^k is a *contravariant functor* (see, for example, [J: *Top*], pp. 69 and 66) from the category of real vector spaces and linear maps to itself. We now state this in detail.

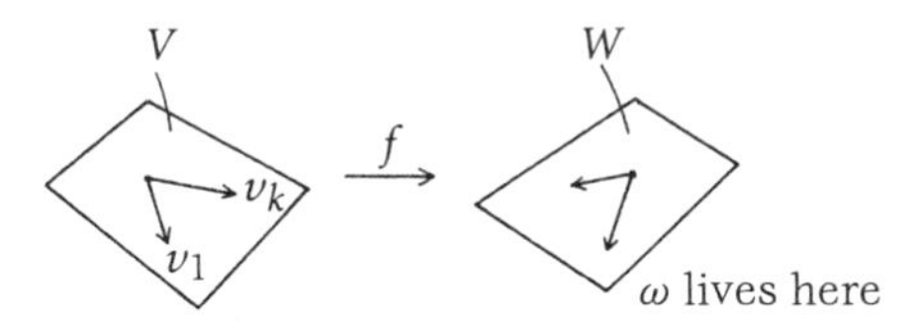

Figure 3.1. Defining $\mathrm{Alt}^k f$

Definition and Note. *If $f : V \to W$ is linear, then the linear map*

$$\mathrm{Alt}^k f : \mathrm{Alt}^k W \longrightarrow \mathrm{Alt}^k V$$

is defined by $((\mathrm{Alt}^k f)(\omega))(v_1, \ldots, v_k) := \omega(f(v_1), \ldots, f(v_k))$ *and by the convention* $\mathrm{Alt}^0 f = \mathrm{Id}_{\mathbb{R}}$. *We have* $\mathrm{Id} \mapsto \mathrm{Id}$ *and the contravariant chain rule:*

$$\mathrm{Alt}^k \mathrm{Id}_V = \mathrm{Id}_{\mathrm{Alt}^k V} \quad \textit{and}$$

$$\mathrm{Alt}^k(g \circ f) = \mathrm{Alt}^k f \circ \mathrm{Alt}^k g$$

for linear maps $V \xrightarrow{f} W \xrightarrow{g} X$.

A great many functors are used in mathematics, and in case of doubt applying the notation for each functor (here Alt^k) to the corresponding morphisms makes things nice and clear. But not all cases are doubtful, and in practice one makes do, for hundreds of functors, with *two* ways of writing the morphisms corresponding to a given f, namely f_* in the covariant and f^* in the contravariant case. This is not only convenient but also informative, so when there is no risk of confusion we adopt this convention here.

Notation and terminology. Instead of $\mathrm{Alt}^k f$ we just write f^*, and call $f^*\omega$ the k-form ***induced*** by f from ω.

3.2 The Components of an Alternating k-Form

We must also know how to compute with alternating k-forms in terms of a basis, since later we will sometimes have to consider differential forms on manifolds in local coordinates. If a basis of V is chosen, then an alternating k-form, like any multilinear form, can be characterized by the numbers it assigns to (k-tuples of) basis vectors.

Terminology. If $(e_1, \dots, e_n)$ is a basis of V and ω an alternating k-form on V, then the numbers

$$a_{\mu_1 \dots \mu_k} := \omega(e_{\mu_1}, \dots, e_{\mu_k}),$$

for $1 \le \mu_i \le n$, are called the ***components*** of ω with respect to the basis.

Because ω is alternating, the components are of course "skew symmetric" in their indices; that is,

$$a_{\mu_{\tau(1)} \dots \mu_{\tau(k)}} = \operatorname{sgn}(\tau) a_{\mu_1 \dots \mu_k}.$$

Hence it suffices to know $a_{\mu_1 \dots \mu_k}$ for $\mu_1 < \dots < \mu_k$. But there are no further relations among the components: the $a_{\mu_1 \dots \mu_k}$ can be prescribed arbitrarily for $\mu_1 < \dots < \mu_k$. The next lemma makes this precise.

Lemma. *If $(e_1, \dots, e_n)$ is a basis of V, then the map*

$$\begin{aligned} \operatorname{Alt}^k V &\longrightarrow \mathbb{R}^{\binom{n}{k}}, \\ \omega &\longmapsto (\omega(e_{\mu_1}, \dots, e_{\mu_k}))_{\mu_1 < \dots < \mu_k} \end{aligned}$$

is an isomorphism.

Proof. The map is clearly linear. Because ω is multilinear, we always have

$$\omega\left(\sum_{\mu_1} v_{(1)}^{\mu_1} e_{\mu_1}, \dots, \sum_{\mu_k} v_{(k)}^{\mu_k} e_{\mu_k}\right) = \sum_{\mu_1, \dots, \mu_k} v_{(1)}^{\mu_1} \cdot \ldots \cdot v_{(k)}^{\mu_k} \omega(e_{\mu_1}, \dots, e_{\mu_k}).$$

If $\omega(e_{\mu_1}, \dots, e_{\mu_k}) = 0$ for $\mu_1 < \dots < \mu_k$, then the same holds for all other $\mu_1, \dots, \mu_k$ since ω is alternating. Hence the map $\operatorname{Alt}^k V \to \mathbb{R}^{\binom{n}{k}}$ is injective.

But the map is also surjective. To show this, let

$$(a_{\mu_1 \dots \mu_k})_{\mu_1 < \dots < \mu_k} \in \mathbb{R}^{\binom{n}{k}}$$

be arbitrarily prescribed. For arbitrary indices, we define

$$a_{\mu_1\ldots\mu_k} := \begin{cases} 0 & \text{if two of the indices agree,} \\ \operatorname{sgn}(\tau) a_{\mu_{\tau(1)}\ldots\mu_{\tau(k)}} & \text{otherwise,} \end{cases}$$

where each $\tau : \{1, \ldots, k\} \to \{1, \ldots, k\}$ is the permutation that orders the indices by magnitude: $\mu_{\tau(1)} < \cdots < \mu_{\tau(k)}$. Then the desired alternating k-form is given by

$$\omega(v_1, \ldots, v_k) := \sum_{\mu_1, \ldots, \mu_k} v_{(1)}^{\mu_1} \cdot \ldots \cdot v_{(k)}^{\mu_k} a_{\mu_1\ldots\mu_k},$$

where of course $v_{(j)}^1, \ldots, v_{(j)}^n$ denote the components of $v_j \in V$ with respect to $(e_1, \ldots, e_n)$. □

Corollary. *If* $\dim V = n$, *then* $\dim \operatorname{Alt}^k V = \binom{n}{k}$.

For $k = 0$ this agrees with the convention $\operatorname{Alt}^0 V := \mathbb{R}$, and for $k = 1$ it is the well-known fact that $\dim V^* = \dim V$. But the dimension of $\operatorname{Alt}^{n-1} V$ is also n, so the alternating $(n-1)$-forms, the 1-forms, and the elements ("vectors") of V itself are all represented in computations in coordinates by n-tuples of real numbers. Despite this, they should not be confused with each other, because each of the n-tuples behaves differently in passing from one basis to another. Vectors, 1-forms, and alternating $(n-1)$-forms are just *not* canonically the same, and when you use isomorphisms

$$V \cong V^* \cong \operatorname{Alt}^{n-1} V,$$

which because of the equality of dimensions is of course possible and sometimes useful, you have to keep in mind that such isomorphisms are not given canonically but rather *chosen*. (An isomorphism $\varphi : V \cong V^*$ corresponds to the choice of a nondegenerate bilinear form β on $V \times V$, namely $\varphi(v)(w) = \beta(v, w)$; an isomorphism $V \cong \operatorname{Alt}^{n-1} V$ corresponds to the choice of a basis element in $\operatorname{Alt}^n V$. See Exercise 3.1.)

3.3 Alternating n-Forms and the Determinant

Of particular interest for integration theory on manifolds are the alternating n-forms, where $n = \dim V$. What we know about them so far is that $\dim \mathrm{Alt}^n V = 1$. We can also write this as follows:

Corollary. *If $(e_1, \ldots, e_n)$ is a basis of V and $a \in \mathbb{R}$, then there is exactly one alternating n-form ω on V such that*

$$\omega(e_1, \ldots, e_n) = a.$$

When $(e_1, \ldots, e_n)$ is the standard basis of $\mathbb{R}^n$ and $a = 1$, this is the *determinant* $\det : M(n \times n, \mathbb{R}) \to \mathbb{R}$, viewed as a multilinear form on the column vectors, as you know from linear algebra. The determinant is the only map from the space of $n \times n$ matrices over $\mathbb{K}$ into $\mathbb{K}$ that is multilinear and alternating on the columns and assigns the value $1 \in \mathbb{K}$ to the identity matrix. For arbitrary endomorphisms $f : V \to V$, we have the following result.

Lemma. *If V is an n-dimensional real vector space and $f : V \to V$ is linear, then $\mathrm{Alt}^n f : \mathrm{Alt}^n V \to \mathrm{Alt}^n V$ is multiplication by $\det f \in \mathbb{R}$.*

Proof. Note that since $\mathrm{Alt}^n V = 1$ the statement could also serve as a coordinate-free *definition* of $\det f$. But we already know $\det f$ according to the usual definition $\det f := \det(\varphi^{-1} \circ f \circ \varphi)$ for some (hence every) $\varphi : \mathbb{R}^n \cong V$:

$$\begin{array}{ccc} V & \xrightarrow{f} & V \\ \cong \uparrow \varphi & & \cong \uparrow \varphi \\ \mathbb{R}^n & \xrightarrow{A} & \mathbb{R}^n \end{array}$$

Thus $\det f = \det A$, and the lemma has content and needs a proof.

The following diagram is commutative by the chain rule for the functor Alt^n:

$$\begin{array}{ccc} \mathrm{Alt}^n V & \xleftarrow{\mathrm{Alt}^n f} & \mathrm{Alt}^n V \\ {\scriptstyle \mathrm{Alt}^n\varphi}\downarrow\cong & & {\scriptstyle \mathrm{Alt}^n\varphi}\downarrow\cong \\ \mathrm{Alt}^n \mathbb{R}^n & \xleftarrow{\mathrm{Alt}^n A} & \mathrm{Alt}^n \mathbb{R}^n \end{array}$$

Therefore $\mathrm{Alt}^n f$ and $\mathrm{Alt}^n A$ are given by multiplication by one and the same real number. To determine it we apply $\mathrm{Alt}^n A$ to the element $\det \in \mathrm{Alt}^n \mathbb{R}^n$, and for the canonical basis $(e_1, \ldots, e_n)$ of $\mathbb{R}^n$ we obtain

$$\begin{aligned} ((\mathrm{Alt}^n A)(\det))(e_1, \ldots, e_n) &= \det(Ae_1, \ldots, Ae_n) \\ &= \det A \\ &= \det A \cdot \det(e_1, \ldots, e_n). \end{aligned}$$

Thus $\det A = \det f$ is the desired factor. □

Finally, let me point out explicitly that $n+1$ vectors in an n-dimensional vector space V are always linearly dependent. Hence any alternating k-form on V must vanish for $k > n$, and this is confirmed by the dimension formula $\mathrm{Alt}^k V = \binom{n}{k}$.

Note. $\mathrm{Alt}^k V = 0$ *for* $k > \dim V$.

3.4 Differential Forms

Now let's climb from the flatlands of linear algebra to the rolling hills of manifolds!

Definition. A ***differential form of degree k***, or simply a ***k-form***, on a manifold M is a correspondence ω that assigns to every $p \in M$ an alternating k-form $\omega_p \in \mathrm{Alt}^k T_p M$ on the tangent space at p.

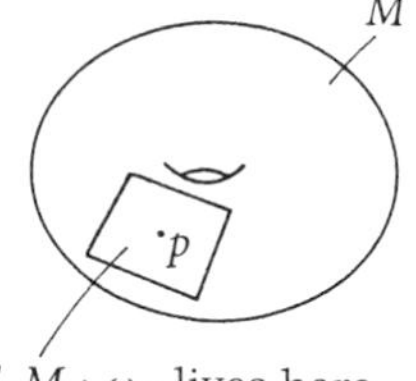

Figure 3.2. A k-form ω on M: the assignment $p \mapsto \omega_p \in \mathrm{Alt}^k T_p M$.

A differential form assigns an ω_p to each $p \in M$, and ω_p in turn assigns a number to each k-tuple of vectors in $T_p M$. That's a lot of assigning! To aid our intuition, let's think of the many ω_p's not as busily assigning but as quietly sitting there waiting. Only when called on by vectors will they respond with a number.

We denote the component functions of a k-form ω on M relative to a chart (U, h) by

$$\omega_{\mu_1 \dots \mu_k} := \omega(\partial_{\mu_1}, \dots, \partial_{\mu_k}) : U \to \mathbb{R},$$

and of course we call a k-form ***continuous***, ***differentiable***, etc. if its component functions relative to the charts of some (hence every) atlas in the differentiable structure on M are continuous, differentiable, etc.

Keep firmly in mind that according to our interpretation (given in detail in Section 2.8) of $\partial_\mu = \partial/\partial x^\mu$ as the canonical basis vector fields of the chart, the component functions $\omega_{\mu_1 \dots \mu_k}$ are really defined "upstairs" on $U \subset M$. Of course, we can also use h to "pull them down," but then they become $\omega_{\mu_1 \dots \mu_k} \circ h^{-1}$.

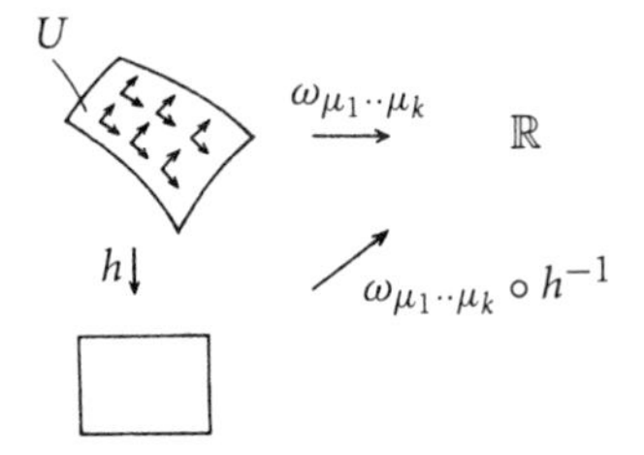

Figure 3.3. Component functions live "upstairs."

Two more remarks on terminology. First, the word "alternating" has somehow worn off in differential calculus, and one speaks simply of differential forms or k-forms ω on M. But the individual $\omega_p : T_pM \times \cdots \times T_pM \to \mathbb{R}$ are always understood to be alternating, and of course the definition above says so. Second, for the present we don't want to restrict our attention exclusively to *differentiable* k-forms because we begin by dealing only with the integration of k-forms on k-dimensional manifolds, and in this situation differentiability would be an unnecessarily strong demand to make on ω. So for the time being we have to add the word "differentiable" whenever we mean it. Later, though, the differentiable differential forms will again be the central objects of interest, and we introduce the usual notation now.

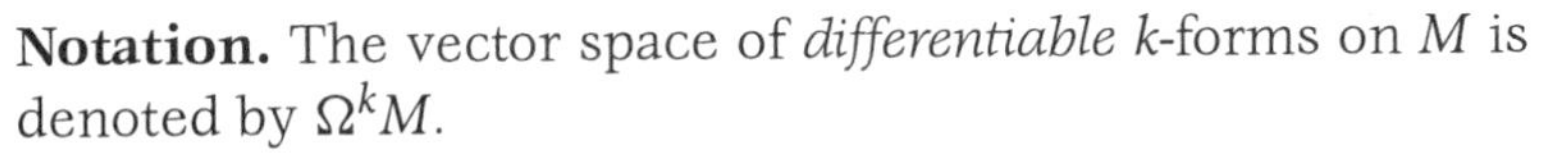

Notation. The vector space of *differentiable* k-forms on M is denoted by $\Omega^k M$.

Since $\mathrm{Alt}^0 T_pM = \mathbb{R}$, we have $\Omega^0 M = C^\infty(M)$, the ring of differentiable functions on M, or at any rate this is how we intend the differentiability of 0-forms to be understood. A 0-form $\omega : M \to \mathbb{R}$ is just its own (and its only) component function; it has $k = 0$ indices, so none at all.

A differentiable map $f : M \to N$ canonically induces a linear map from $\Omega^k N$ to $\Omega^k M$, which we write, again using the standard notation, as

$$f^* : \Omega^k N \to \Omega^k M.$$

To be precise, $f^*\omega$ is defined for $\omega \in \Omega^k N$ and $v_1, \ldots, v_k \in T_p M$ by

$$(f^*\omega)_p(v_1, \ldots, v_k) := \omega_{f(p)}(df_p v_1, \ldots, df_p v_k)$$

—in what other obvious way could $\omega \in \Omega^k N$ respond through f to vectors $v_1, \ldots, v_k \in T_p M$? The correspondence $f^* =: \Omega^k f : \Omega^k N \to \Omega^k M$ is thus given "pointwise" (i.e. for each $p \in M$) by $\mathrm{Alt}^k(df_p)$.

But since the differential and Alt^k are both functorial ("$\mathrm{Id} \mapsto \mathrm{Id}$ and the chain rule"), we also have the following result.

Note. *Ω^k canonically gives a contravariant functor from the differential category to the linear category. In other words, if $f^* : \Omega^k N \to \Omega^k M$ denotes the linear map induced by a differentiable function $f : M \to N$, then $(\mathrm{Id}_M)^* = \mathrm{Id}_{\Omega^k M}$ and $(g \circ f)^* = f^* \circ g^*$.*

3.5 One-Forms

The differentiable 1-forms, the $\omega \in \Omega^1 M$, are also called ***Pfaffian forms***. The differentials of differentiable functions are a particular kind of Pfaffian forms ("exact Pfaffian forms").

Definition. Let $f : M \to \mathbb{R}$ be a differentiable function. Then the differentiable 1-form $df \in \Omega^1 M$ given by $p \mapsto df_p \in \mathrm{Alt}^1 T_p M$ is called the ***differential*** of f.

The differential df_p at the single point $p \in M$ would actually be a linear map $df_p : T_p M \to T_{f(p)}\mathbb{R}$, but of course we are referring to the *canonical* isomorphism $\mathbb{R} \cong T_{f(p)}\mathbb{R}$ (see Section 2.6) and thinking of df_p as an element of the dual space $T_p^* M$ of $T_p M$. In this sense,we also have $df_p(v) = v(f)$

for $v \in T_pM$, because for instance $df_p(\dot{\alpha}(0)) = (f \circ \alpha)^{\cdot}(0) \in \mathbb{R}$ (see Section 2.7). So in local coordinates, i.e. relative to a chart (U, h), the n component functions of df are just

$$df(\partial_\mu) = \partial_\mu f, \quad \mu = 1, \ldots, n.$$

Exercise 2.3 already dealt with the fact that the n-tuple $(\partial_1 f, \ldots, \partial_n f)$ does not define a tangent vector $\mathcal{D}_pM \to \mathbb{R}^n$. Now we see what is, from our present viewpoint, the "true" meaning of partial derivatives with respect to coordinates: They are the components of the differential df, which thus assumes the role of the gradient on manifolds.

In particular, for a chart $h = (x^1, \ldots, x^n)$ on U, we can take the differentials $dx^\mu \in \Omega^1 U$ *of the coordinate functions* x^μ *themselves*. Their components $dx^\mu(\partial_\nu)$, $\nu = 1, \ldots, n$, are

$$dx^\mu(\partial_\nu) = \partial_\nu x^\mu = \delta^\mu_\nu := \begin{cases} 1 & \text{if } \mu = \nu, \\ 0 & \text{if } \mu \neq \nu. \end{cases}$$

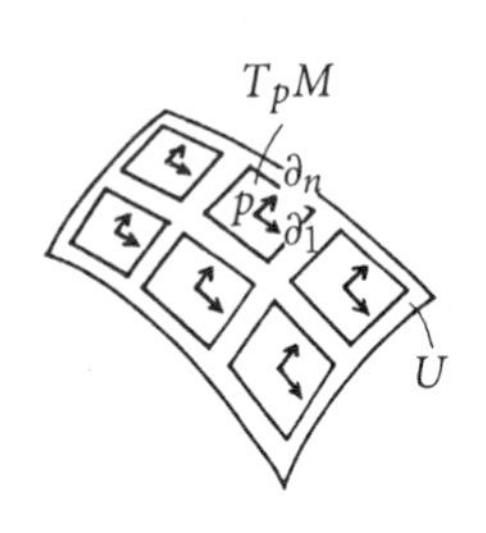

Figure 3.4. $(dx^1, \ldots, dx^n)$ is dual to $(\partial_1, \ldots, \partial_n)$ everywhere on U.

Lemma. *At each point $p \in U$, the basis of T_p^*M dual to $(\partial_1, \ldots, \partial_n)$ is $(dx^1_p, \ldots, dx^n_p)$, where $dx^1, \ldots, dx^n \in \Omega^1 U$ are the differentials of the coordinate functions $x^\mu : U \to \mathbb{R}$ of a chart.*

Corollary. *If (U, h) is a chart on M, where $h = (x^1, \ldots, x^n)$, and ω is a 1-form, then*

$$\omega | U = \sum_{\mu=1}^{n} \omega_\mu dx^\mu.$$

Here $\omega_\mu : U \to \mathbb{R}$ are the component functions $\omega_\mu := \omega(\partial_\mu)$. In particular, for differentiable functions we have

$$df = \sum_{\mu=1}^{n} \frac{\partial f}{\partial x^\mu} dx^\mu$$

on the chart domain U.

Proof. We test equality at each point $p \in U$ by inserting the basis vectors ∂_ν, $\nu = 1, \ldots, n$, on both sides: $\omega_p(\partial_\nu) = \omega_\nu(p)$

by the definition of ω_ν, and

$$\sum_{\nu=1}^{n} \omega_\mu(p) dx_p^\mu(\partial_\nu) = \sum_{\mu=1}^{n} \omega_\mu(p)\delta_\nu^\mu = \omega_\nu(p).$$

Hence the two 1-forms are equal on U. □

This local description of 1-forms as $\omega = \sum_{\mu=1}^{n} \omega_\mu dx^\mu$, and in particular of differentials as $df = \sum_{\mu=1}^{n} \partial_\mu f \cdot dx^\mu$, is the key to computing with these forms in coordinates. It is used quite often for local concepts and proofs. But such a description is possible for more than just 1-forms. Once the *exterior product* or *wedge product* has been introduced, a k-form will be expressible relative to a chart as

$$\omega = \sum_{\mu_1 < \cdots < \mu_k} \omega_{\mu_1 \ldots \mu_k} dx^{\mu_1} \wedge \cdots \wedge dx^{\mu_k}$$

in terms of component functions and differentials of the coordinates, and local computations with k-forms will reduce to computations with familiar *functions*.

3.6 Test

(1) Let $f_i, g_i : V \to \mathbb{R}$ be linear maps. Then the map $V \times \cdots \times V \to \mathbb{R}$ given by $(v_1, \ldots, v_k) \mapsto$

- □ $f_1(v_1) \cdot \ldots \cdot f_k(v_k) + g_1(v_1) \cdot \ldots \cdot g_k(v_k)$
- □ $f_1(v_1) + \cdots + f_k(v_k) + g_1(v_1) + \cdots + g_k(v_k)$
- □ $(f_1(v_1) + g_1(v_1)) \cdot \ldots \cdot (f_k(v_k) + g_k(v_k))$

is multilinear.

(2) Let $f : V \times \cdots \times V \to \mathbb{R}$ be a multilinear map. Which of the following conditions are sufficient for f to be alternating?

- □ $f(v_1, \ldots, v_k) = 0$ as soon as $v_i = v_{i+1}$ for some i.
- □ There exists an $\varepsilon : \mathcal{S}_n \to \{-1, +1\}$, not the constant $+1$, such that $f(v_{\tau(1)}, \ldots, v_{\tau(k)}) = \varepsilon(\tau) f(v_1, \ldots, v_k)$.
- □ $f(v, \ldots, v) = 0$ for all $v \in V$.

(3) Let $\mathrm{Alt}^k(V, W)$ denote the vector space of alternating k-linear maps from $V \times \cdots \times V$ to W. If $\dim V = n$ and $\dim W = m$, then $\dim \mathrm{Alt}^k(V, W) =$

☐ $\binom{m+n}{k}$. ☐ $m + \binom{n}{k}$. ☐ $m\binom{n}{k}$.

(4) Does the cross product of vectors in $\mathbb{R}^3$ define an element of the space $\mathrm{Alt}^2(\mathbb{R}^3, \mathbb{R}^3)$?

☐ Yes, because the cross product is bilinear and skew symmetric.

☐ No, because the cross product is skew symmetric but not alternating.

☐ No, because the cross product is linear, not bilinear.

(5) Let V be an n-dimensional vector space, $k > 0$, ω an alternating k-form on V, and $\overline{v}_i = \sum_{j=1}^{k} a_{ij} v_j$. For what values of k do we have $\omega(\overline{v}_1, \ldots, \overline{v}_k) = \det a \cdot \omega(v_1, \ldots, v_k)$?

☐ only for $k = n$.

☐ only for $k = 1$ and $k = n$.

☐ for all k.

(6) Let M be a nonempty manifold with $\dim M = n > 0$ and $0 \leq k \leq n$. Then $\dim \Omega^k M =$

☐ ∞. ☐ $\binom{n}{k}$. ☐ $k(k-1)/2$.

(7) Let $f : M \to S^1 \subset \mathbb{C}$ be a differentiable function. We can always write $f = e^{i\theta}$, but in general the function θ will not be continuous, let alone differentiable. Locally, however, θ is well defined as a differentiable real-valued function up to addition of an integer multiple of 2π. Hence $\sin\theta$, $\cos\theta \in \Omega^0 M$ and $d\theta \in \Omega^1 M$ are well defined. Moreover, as a complex-valued function, f also has a complex-valued differential $df \in \Omega^1(M, \mathbb{C})$. We have

☐ $df = e^{id\theta}$.

☐ $df = -\sin\theta \, d\theta + i \cos\theta \, d\theta$.

☐ $df = if\,d\theta$.

(8) Let $M \neq \emptyset$ and $1 \leq k \leq n = \dim M$. Can there exist a map $f : M \to M$ with the property that $f^*\omega = -\omega$ for all $\omega \in \Omega^k M$?

☐ Yes, for instance with $M = \mathbb{R}^n$, $f(x) := -x$, and k odd.

☐ Yes, for instance with $M := S^n$, f the antipodal map, and k arbitrary.

☐ No, never.

(9) Let $\pi : \mathbb{R}^2 \setminus \{0\} \to S^1$ be radial projection and η a 1-form on S^1. We consider the tangent vector $v := \binom{0}{1} \in \mathbb{R}^2 \cong T_p(\mathbb{R}^2 \setminus \{0\})$ at the point $p \in \mathbb{R}^2 \setminus \{0\}$, and similarly $w := \binom{0}{r} \in \mathbb{R}^2 \cong T_{rp}(\mathbb{R}^2 \setminus \{0\})$ at the point rp for $r > 0$. Which of the following is true?

☐ $\pi^*\eta(w) = \pi^*\eta(v)$.

☐ $\pi^*\eta(w) = r\pi^*\eta(v)$.

☐ $r\pi^*\eta(w) = \pi^*\eta(v)$.

(10) Now let π be the radial projection of $\mathbb{R}^3 \setminus \{0\}$ onto S^2 and $\iota : S^2 \hookrightarrow \mathbb{R}^3 \setminus \{0\}$ the inclusion map. Let $\eta \in \Omega^3(\mathbb{R}^3 \setminus \{0\})$ and $\omega \in \Omega^2 S^2$. Then

☐ $\pi^*\iota^*\eta = \eta$. ☐ $\pi^*\iota^*\eta = 0$. ☐ $\iota^*\pi^*\omega = \omega$.

3.7 Exercises

EXERCISE 3.1. Let V be an n-dimensional vector space and let $\omega \in \mathrm{Alt}^n V$ be nonzero. Show that the map

$$\begin{array}{ccc} V & \longrightarrow & \mathrm{Alt}^{n-1} V, \\ v & \longmapsto & v \lrcorner\, \omega, \end{array}$$

where $(v \lrcorner\, \omega)(v_1, \ldots, v_{n-1}) := \omega(v, v_1, \ldots, v_{n-1})$, is an isomorphism.

EXERCISE 3.2. Let $(e_1, \ldots, e_n)$ be an orthonormal basis in the Euclidean vector space $(V, \langle \cdot, \cdot \rangle)$, and let ω be the alternating n-form on V with $\omega(e_1, \ldots, e_n) = 1$. Compute the "density" $|\omega(v_1, \ldots, v_n)|$ from the "first fundamental form" $(g_{\mu\nu})_{\mu,\nu=1,\ldots,n}$, where $g_{\mu\nu} := \langle v_\mu, v_\nu \rangle$.

EXERCISE 3.3. Determine the transformation formula for k-forms in the Ricci calculus. More precisely, for charts (U, h) and $(U, \overline{h})$, write the coordinates as

$$h = (x^1, \ldots, x^n) \quad \text{and}$$
$$\overline{h} = (x^{\overline{1}}, \ldots, x^{\overline{n}}),$$

and write the component functions of $\omega \in \Omega^k M$ with respect to the coordinates accordingly. How can the $\omega_{\overline{\mu}_1 \ldots \overline{\mu}_k}$ be computed from the $\omega_{\mu_1 \ldots \mu_k}$?

EXERCISE 3.4. If $V_\alpha^+ \subset \mathbb{R}^2$ is the closed ray with initial point 0 and angle α to the positive x-axis, the angle function

$$\varphi_\alpha : \mathbb{R}^2 \setminus V_\alpha^+ \longrightarrow (\alpha - 2\pi, \alpha)$$

of the polar coordinates is a well-defined differentiable function. Let its differential be denoted by $\omega_\alpha := d\varphi_\alpha$. Then any two differentials ω_α and ω_β agree on $\mathbb{R}^2 \setminus (V_\alpha^+ \cup V_\beta^+)$ (why?), so the ω_α's unambiguously define a Pfaffian form $\omega \in \Omega^1(\mathbb{R}^2 \setminus \{0\})$. This is a popular model for certain phenomena. Prove that there is no differentiable function $f : \mathbb{R}^2 \setminus \{0\} \to \mathbb{R}$ such that $\omega = df$.

3.8 Hints for the Exercises

FOR EXERCISE 3.1. I suggest reading and saying the expression $v \lrcorner \omega$ as "v in ω," because this will remind us of the meaning of the symbol $\lrcorner : v \lrcorner \omega = \omega(v, \ldots)$. Since V and $\text{Alt}^{n-1} V$ are often identified—not to say confused—it's useful to clarify in your own mind what role is played by the choice of an n-form ω. Incidentally, even for a given ω the map can't quite be called canonical, because v could equally well be inserted as the last variable in ω and this would change the sign of the map by $(-1)^{n-1}$. But from now on we'll stick with the definition given in this exercise. Considered technically, the exercise is straightforward, and I don't know what other hints I could give.

FOR EXERCISE 3.2. The formula you are asked to find and prove here plays an important role in integration in local coordinates on "Riemannian" manifolds, in particular on submanifolds of $\mathbb{R}^n$. Instead of a fixed vector space V, one then deals with tangent spaces to a chart domain, and the $v_1, \ldots, v_n$ are the $\partial_1, \ldots, \partial_n$.

In principle, these functions $g_{\mu\nu} : U \to \mathbb{R}$ are easy to compute. But to integrate in this situation, one needs the function $|\omega(\partial_1, \ldots, \partial_n)| : U \to \mathbb{R}_+$. (Of course, part of the solution to Exercise 3.2 is that $|\omega|$ is independent of the choice of orthonormal basis, but this is also easy to show directly: isometric transformations take orthonormal bases to orthonormal bases and always have determinant ± 1 ....)

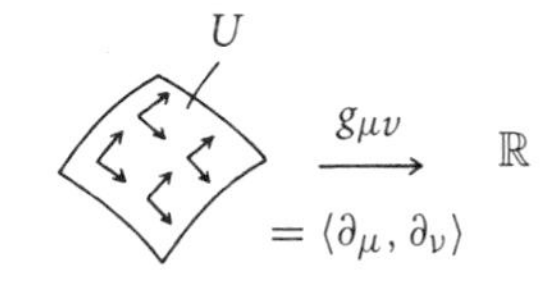

Figure 3.5. "Components of the first fundamental form"

This is the deeper meaning of the problem! Superficially, it's a useful exercise in working with n-forms, matrices, scalar products, how n-forms behave under a change of coordinates, etc. A practical hint: First figure out the relationship between the matrix $G := (g_{\mu\nu})$ and the matrix $A = (a_{\mu\nu})$ that describes the expansion of the v_μ in terms of the orthonormal basis $e_1, \ldots, e_n$, i.e. the one satisfying $v_\mu =: \sum_{\nu=1}^n a_{\mu\nu} e_\nu$.

FOR EXERCISE 3.3. As you see, the intersection of two chart domains is already denoted here without loss of generality by U; otherwise we would have had to consider $U \cap V$. I probably don't have to argue that the question about the transformation formula for the components of a k-form is meaningful and legitimate. But in addition to this useful information, the exercise also gives you the opportunity to learn some really elegant notation from the Ricci calculus's bag of tricks. You just have to be able to read it! You can see, of course, that the notation $\omega_{\mu_1 \ldots \mu_k} := \omega(\partial_{\mu_1}, \ldots, \partial_{\mu_k})$ for the components of a k-form says nothing about the coordinates being used—in complete accord with the Ricci philosophy that the coordinates themselves aren't given individual names. And how awkward any other approach would be! But now, what if a second coordinate system has to be considered? Answer: Put bars over—the indices! This doesn't just create new names for indices (as it would be read without a more detailed explanation), it also means that the quantities with barred indices refer to the second coordinate system. Just try using this notation. It works great!

FOR EXERCISE 3.4. You also know the "argument" function

$$\varphi_\alpha : \mathbb{R}^2 \setminus V_\alpha^+ \to (\alpha - 2\pi, \alpha)$$

from complex function theory; for $\alpha = \pi$, for instance, it's the imaginary part of the principal branch of the logarithm. Not directly, but in spirit, our exercise is related to the fact that although $d(\ln z)/dz = 1/z$ is defined on all of $\mathbb{C} \setminus \{0\}$, it has no antiderivative there.

The problem isn't hard. What could be said about $f - \varphi_\pi$ (for example) if there did exist such an f? And would that be possible?

Thus a Pfaffian form can be *locally* a differential everywhere without having to be one *globally*. This is a mathematically important phenomenon ("de Rham cohomology"), and the example given in the exercise may be the simplest one there is: no wonder it's used so often. You should know it. In physics, it plays a role in interpreting the Aharonov-Bohm effect.

4 CHAPTER The Concept of Orientation

4.1 Introduction

As you know, the *direction* of integration matters when you integrate a function of a real variable:

$$\int_a^b f(x)\,dx = -\int_b^a f(x)\,dx.$$

The dx senses, so to speak, when the direction of integration is reversed: the differences $\Delta x_k = x_{k+1} - x_k$ in the Riemann sums $\sum f(x_k)\Delta x_k$ are positive or negative according to whether the partition points are increasing or decreasing. The same thing happens with line integrals

$$\int_\gamma f(x,y,z)\,dx + g(x,y,z)\,dy + h(x,y,z)\,dz,$$

where γ is a curve in $\mathbb{R}^3$, and with contour integrals $\int_\gamma f(z)\,dz$ in complex function theory. They are invariant under all reparametrizations of the curve that do not change the direction in which the curve is traced. But if the curve is traced backwards, the sign of the integral is reversed.

I don't mean to say that this response to a change of direction is necessarily a property of every meaningful version of integration. For example, the *arc length* $\int_\gamma ds$ of a curve should be independent of the direction in which the curve is traced, and in fact the so-called line element

$$ds = \sqrt{dx^2 + dy^2 + dz^2}$$

(not a 1-form!) doesn't sense a reversal of direction. But we usually deal with integrals that are sensitive to direction, and for the setup of vector analysis it is necessary, for this and other reasons, to generalize the concept of a directed interval to that of an *oriented manifold.* As a preliminary step we need the linear-algebraic version, namely the concept of an oriented n-dimensional real vector space.

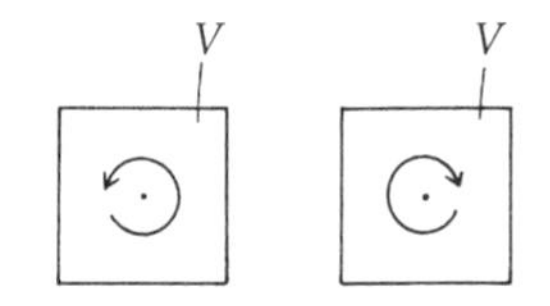

Figure 4.1. The two orientations of a two-dimensional real vector space

In order to get a first intuitive idea of orientation, we consider the dimensions $n = 1, 2,$ and 3, which are directly accessible to our intuition. To "orient" a one-dimensional vector space means to choose a *direction* in the space, and it is intuitively clear that this is possible in exactly two different ways. To orient a two-dimensional real vector space V, we must define one of the two *senses of rotation* in V to be positive. Of course, as long as an exact mathematical definition isn't required, everyone has a perfectly good intuitive idea what a "sense of rotation" is, and a fair number of people will at least have heard that the "mathematically positive" sense of rotation is counterclockwise. So it may not be completely unnecessary to point out that in a two-dimensional vector space V there is no well-defined "clockwise sense." The mathematically positive sense of rotation can't be specified unless V already *is* oriented.

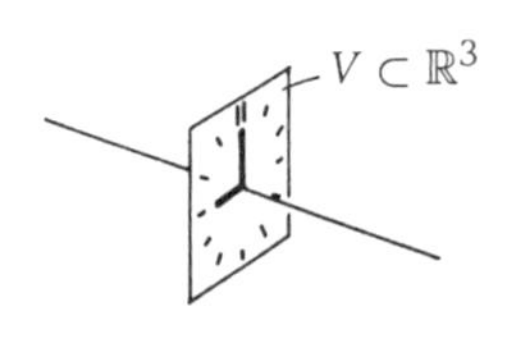

Figure 4.2. On the "transparent" two-dimensional clock, is it 9:00 or 3:00?

Finally, in a three-dimensional real vector space, the purpose of an orientation is to distinguish a "screw sense," or to determine what "right-handedness" should mean. This expression refers to the familiar *right-hand rule*, which says that a basis (v_1, v_2, v_3) is called "right-handed" if the three vectors in this order correspond to the directions of the thumb, index finger, and middle finger of a *right* hand. It takes a certain effort to free oneself from the illusion that the right-hand rule actually orients all three-dimensional vector spaces. But once we start to think about it, we soon realize that we can intuitively compare the position of three vectors in a three-dimensional vector space V with our right hand only after mapping V onto the real physical space that surrounds us, and whether (v_1, v_2, v_3) turns out to be right-handed or left-handed depends on how we do that: in a mirror, a right hand looks like a left.

4.2 The Two Orientations of an n-Dimensional Real Vector Space

But how can orientation be understood precisely as a mathematical concept? There are several equivalent possibilities. Our approach is based on taking as definition a version that can't immediately be pictured but is easy to work with. First we assume that $\dim V > 0$.

Definition. Two bases $(v_1, \ldots, v_n)$ and $(w_1, \ldots, w_n)$ of a real vector space V are said to ***have the same orientation***, written

$$(v_1, \ldots, v_n) \sim (w_1, \ldots, w_n),$$

if one basis is mapped to the other by a transformation with positive determinant, that is, if $\det f > 0$ holds for the automorphism $f : V \to V$ with $f(v_i) = w_i$.

Note and Definition. *The property of having the same orientation is clearly an equivalence relation on the set $\mathfrak{B}(V)$ of bases of V, and it has exactly two equivalence classes.* These equivalence classes are called the two ***orientations*** of V: An ***oriented vector space*** is a pair $(V, \mathcal{O})$ consisting of a finite-dimensional real vector space V and one of its two orientations.

Up to now we have assumed that V is positive-dimensional. If we were to take the definition literally for zero-dimensional spaces, they would be canonically oriented, since $\{0\}$ has only the empty basis and hence only one equivalence class of bases with the same orientation. But introducing another "orientation" for the zero-dimensional spaces, opposite to this canonical one, turns out to be a useful convention.

Convention. Let the numbers ± 1 be the two ***orientations*** of a zero-dimensional real vector space.

Associated with the notion of orientation are some almost self-explanatory terminology and notation. For example, if $(V, \mathcal{O})$ is a (positive-dimensional) oriented vector space, the bases $(v_1, \ldots, v_n) \in \mathcal{O}$ are called ***positively oriented*** and the others ***negatively oriented***. Of course, the ***usual orientation*** of $\mathbb{R}^n$ means the one in which the canonical basis $(e_1, \ldots, e_n)$ is positively oriented.

The orientation, like other additional structures, is usually suppressed in the notation. An isomorphism $f : V \cong W$ between positive-dimensional oriented vector spaces is called ***orientation-preserving*** if it takes some (hence every) positively oriented basis of V to a positively oriented basis of W. In the zero-dimensional case, we naturally call the (unique) map orientation-preserving only if the two orientations are the same (both $+1$ or both -1).

The following *topological* characterization of the orientations of a real vector space is noteworthy and often useful.

Lemma. *If V is an n-dimensional real vector space, $n \geq 1$, then the two orientations of V are the two path components of $\mathfrak{B}(V) \subset V \times \cdots \times V$, the space of bases of V.*

PROOF. Suppose that two bases $B_0 = (v_1, \ldots, v_n)$ and $B_1 = (w_1, \ldots, w_n)$ have different orientations but can be joined by a continuous path $t \mapsto B_t$ in $\mathfrak{B}(V)$. We denote by $f_t : V \to V$ the isomorphism that takes B_0 to B_t. Then the continuous function $t \mapsto \det f_t$ is positive (namely 1) at the left endpoint $t = 0$ of the interval and, by hypothesis, negative at the right endpoint. By the intermediate value theorem it must therefore have a zero, which contradicts the hypothesis that all the f_t are isomorphisms.

Thus bases with different orientations at least lie in different path components of $\mathfrak{B}(V)$. It remains to show that bases B_0 and B_1 with the same orientation can always be connected by a path in $\mathfrak{B}(V)$. We may assume without loss of generality that $V = \mathbb{R}^n$ and B_1 is the standard basis $(e_1, \ldots, e_n)$. We now apply the Gram-Schmidt orthonormalization process to B_0. This takes B_0 to an orthonormal basis in $2n-1$ steps: normalize a vector/take the next vector orthogonal (to the vectors already constructed)/normalize/take the next one orthogonal/normalize/etc. So far this is just a jump from one basis to another, but simply connecting each pair of way stations by a straight line gives us a continuous zigzag path in $\mathfrak{B}(V)$ from B_0 to an orthonormal basis.

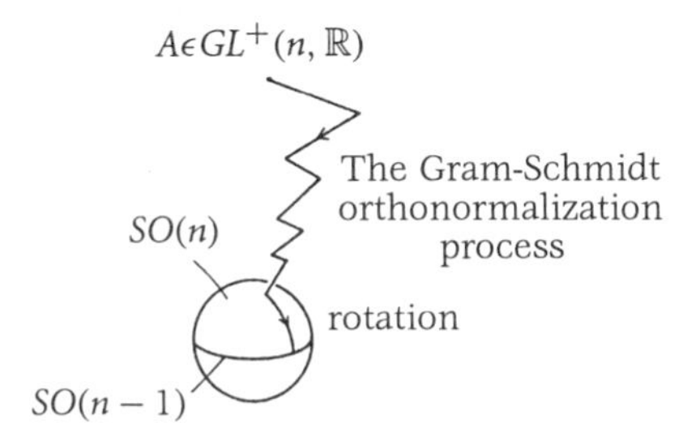

Figure 4.3. Proving the path-connectedness of an orientation

We are left with the problem of how to get from this basis to the standard one along a path in $\mathfrak{B}(V)$. But we can do this along a path that even stays in the space of orthonormal bases. First, by a rotation, we get to an orthonormal basis with first vector e_1. From there, by a rotation in $e_1^\perp$, we reach an orthonormal basis whose first two vectors are e_1 and e_2, etc. After $n-1$ steps our continuous path has brought us to an orthonormal basis $(e_1, \ldots, e_{n-1}, w_n)$, and if there are any difficulties at all they should occur now, because in the one-dimensional space $\{e_1, \ldots, e_{n-1}\}^\perp$ there is no room left to rotate. But now there is no need to rotate, since all three bases have the same orientation:

$$(e_1, \ldots, e_n) \sim (v_1, \ldots, v_n) \sim (e_1, \ldots, e_{n-1}, w_n),$$

the first two by hypothesis and the last because of the path we constructed. So of the two remaining possibilities $w_n = \pm e_n$, only $w_n = e_n$ is possible, and the proof is done. □

4.3 Oriented Manifolds

We orient a manifold by orienting each of its tangent spaces—not in just any way, but in such a way that these orientations get along with each other and don't suddenly "switch." What does this mean? To formulate it precisely, we introduce the following terminology.

Definition. Let M be an n-dimensional manifold. A family $\{\mathcal{O}_p\}_{p\in M}$ of orientations $\mathcal{O}_p$ of its tangent spaces is called ***locally coherent*** if around every point of M there is an ***orientation-preserving chart***, i.e. a chart (U, h) with the property that for every $u \in U$ the differential

$$dh_u : T_uM \xrightarrow{\cong} \mathbb{R}^n$$

takes the orientation $\mathcal{O}_u$ to the usual orientation of $\mathbb{R}^n$.

The simple phrase "locally constant relative to charts" would also have been a reasonable description of this local coherence. In any case, we can now formulate our definition.

Definition. An ***orientation*** of a manifold M is a locally coherent family $\{\mathcal{O}_p\}_{p\in M}$ of orientations of its tangent spaces. An ***oriented manifold*** is a pair $(M, \mathcal{O})$ consisting of a manifold M and an orientation $\mathcal{O}$ of M.

Of course, only for special reasons do we actually denote an oriented manifold by $(M, \mathcal{O})$ rather than simply M.

Definition. A diffeomorphism $f : M \xrightarrow{\cong} \widetilde{M}$ between oriented manifolds is called ***orientation-preserving (resp. orientation-reversing)*** if for every $p \in M$ the differential $df_p : T_pM \xrightarrow{\cong} T_{f(p)}\widetilde{M}$ is orientation-preserving (resp. orientation-reversing).

The usual orientation of $\mathbb{R}^n$ as a vector space makes it into an oriented manifold because canonically $\mathbb{R}^n \cong T_p\mathbb{R}^n$. It is clear that all open (hence full-dimensional) submanifolds of an oriented manifold are also automatically oriented, and in this sense the charts we called "orientation-preserving" at the beginning of this section really are the orientation-preserving charts $h : U \xrightarrow{\cong} U'$ in the sense of the last definition. We make one more observation.

Note. *A chart is orientation-preserving if and only if the basis* $(\partial_1, \ldots, \partial_n)$ *is positively oriented at every point of the chart domain.*

The best training in visualizing the orientation of manifolds is given by the *two-dimensional* manifolds, the ***surfaces***. In intuitive terms, an orientation provides every point of the surface with a sense of rotation, which just indicates which bases for the tangent spaces are positively oriented.

But our intuition about surfaces also immediately reveals a phenomenon that is not immediately obvious in the technical sense, namely the existence of nonorientable manifolds. (See Figure 4.4.) Local coherence, which forbids sudden switches in orientation, is precisely what "obviously" gives contradictory orientation data at the initial point, after a single circuit of the core circle of the Möbius strip. Actually carrying out this argument would of course require that we first *define* the Möbius strip, not just sketch it, then apply Exercise 4.4, which says that a continuous frame field keeps its orientation along a curve in an oriented manifold.

Continuation without switching leads to...

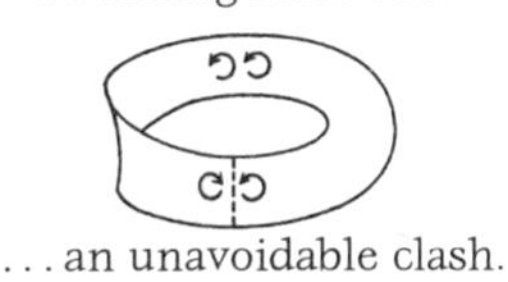

...an unavoidable clash.

Figure 4.4. The Möbius strip, a nonorientable two-dimensional manifold

4.4 Construction of Orientations

It is both intuitively and technically clear that for every orientation of a vector space or a manifold there is also an *opposite orientation*. We introduce notation for it.

Note and Notation. If $\mathcal{O}$ is an orientation of a vector space, let $-\mathcal{O}$ denote the other of the two orientations. *If $\mathcal{O} = \{\mathcal{O}_p\}_{p \in M}$ is an orientation of a manifold M, then so is*

$$-\mathcal{O} = \{-\mathcal{O}_p\}_{p \in M}$$

*(the **opposite** orientation of M).* If the orientation is suppressed in the notation, so that M denotes an oriented manifold, we write $-M$ for the oppositely oriented manifold.

It is also clear that the ***sum*** $M_1 + M_2$ of two oriented n-dimensional manifolds is canonically oriented, simply by $\{\mathcal{O}_p\}_{p \in M_1 + M_2}$. If both summands are nonempty, such a sum thus has at least *four* different orientations, which in the notation just introduced give rise to the four oriented manifolds $\pm M_1 \pm M_2$.

The ***product*** $M_1 \times M_2$ of two oriented manifolds, like their sum, is canonically oriented, but you have to be careful in taking quotients; see Exercise 8.4. Submanifolds of oriented manifolds may not be orientable, as the Möbius strip in $\mathbb{R}^3$ makes clear. But we do have the following result.

Lemma. *Let c be a regular value of a differentiable map $f : M \to N$. If M is orientable, then so is the submanifold $M_0 := f^{-1}(c) \subset M$.*

PROOF. Let orientations for the manifold M and the vector space T_cN be chosen. As we know (see Exercise 2.4), T_pM_0 is the kernel of

$$df_p : T_pM \longrightarrow T_cN.$$

We therefore consider the following linear-algebraic situation: let

$$0 \to V_0 \xrightarrow{\iota} V_1 \xrightarrow{\pi} V_2 \to 0$$

be a "short exact sequence" of linear maps of finite-dimensional real vector spaces. In other words, ι is injective, π is surjective, and $\ker \pi = \operatorname{im} \iota$, as in the case

$$0 \to T_pM_0 \hookrightarrow T_pM \xrightarrow{df_p} T_cN \to 0.$$

Let orientations for V_0, V_1, V_2 be called *compatible* if the following holds: If $v_1, \ldots, v_k$ is a positively oriented basis of V_0 and $\iota(v_1), \ldots, \iota(v_k)$ is extended to a positively oriented basis of V_1 by adjoining $w_1, \ldots, w_{n-k}$, then $\pi(w_1), \ldots, \pi(w_{n-k})$ is a positively oriented basis of V_2. In this sense, for the orientations of any two of the spaces V_0, V_1, V_2 there is exactly one compatible orientation of the third. You can easily convince yourself of this linear-algebraic fact by recalling that for square matrices A and B, every block matrix of the form

$$\begin{pmatrix} A & C \\ & B \end{pmatrix}$$

has determinant $\det A \cdot \det B$. Now if we orient each T_pM_0 compatibly with the orientations of T_pM and T_cN, we obtain a locally coherent family of orientations, hence an orientation of M_0. □

Manifolds can also be oriented by using atlases. To do this, we make a definition.

Definition. An atlas $\mathfrak{A}$ of a differentiable manifold is called an ***orienting atlas*** if all its transition maps w are orientation-preserving, that is, if their Jacobians $\det J_w(x)$ are positive everywhere.

If M is already oriented, the orientation-preserving charts obviously form a maximal orienting atlas. There is a converse.

Note. *If $\mathfrak{A}$ is an orienting atlas of a differentiable manifold M, then there is exactly one orientation of M relative to which all the charts in $\mathfrak{A}$ are orientation-preserving.*

In view of this, we could just as well have defined an orientation as a maximal orienting atlas, and this version of the definition is often preferred because it makes no use of tangent spaces.

4.5 Test

(1) Let $n \geq 1$. When do $(v_1, \ldots, v_n)$ and $(-v_1, \ldots, -v_n)$ have the same orientation?

- ☐ Always.
- ☐ For even n.
- ☐ Never.

(2) How many path components does the orthogonal group $O(n)$ have for $n \geq 3$?

- ☐ One, and this can be shown by using rotations as in the proof of the lemma in Section 4.2.
- ☐ Two, namely $SO(n)$ and $O(n) \setminus SO(n)$.
- ☐ One if n is odd, two if n is even.

(3) Let $\dim V = n$ and $0 \leq k \leq n$. The map $\text{Alt}^k(-\text{Id}_V) : \text{Alt}^k V \to \text{Alt}^k V$ induced by $-\text{Id}_V : V \to V$ is orientation-reversing if and only if the following number is odd:

☐ k. ☐ $\binom{n}{k}$. ☐ $k\binom{n}{k}$.

(4) For diffeomorphisms $f : M \xrightarrow{\cong} N$ between oriented manifolds, the set of x in M for which df_x preserves orientation is

- ☐ open in M, but in general not closed in M.
- ☐ closed in M, but in general not open in M.
- ☐ open and closed in M.

(5) Let M be an oriented manifold. Is a diffeomorphism $f : M \to M$ that is not orientation-preserving necessarily orientation-reversing?

- ☐ Yes, because this is already true for isomorphisms between oriented vector spaces.
- ☐ Yes if M is connected, but otherwise not in general.

☐ No, not even for connected M in general, because df_p can reverse orientation for some p and preserve it for others.

(6) Let M and N be oriented manifolds with dimensions n and k, respectively. Then interchanging the variables defines an *orientation-preserving* diffeomorphism between $N \times M$ and

☐ $M \times N$.

☐ $(-1)^{n \cdot k} M \times N$.

☐ $(-1)^{n+k} M \times N$.

(7) Can a product $M \times N$ of two connected nonempty nonorientable manifolds be orientable?

☐ Yes. For example, $M \times M$ is always orientable.

☐ A product $M \times N$ of nonempty manifolds is orientable if and only if one of the factors is orientable.

☐ A product $M \times N$ of nonempty manifolds is orientable if and only if both the factors are orientable.

(8) Let $\widetilde{M} \to M$ be a covering of n-dimensional manifolds. (This notion hasn't been explained, but it comes up again in Exercise 5.4 and a reference is given there. So you may either ignore this question for now or look ahead at the reference.)

☐ If $\widetilde{M}$ is orientable, then so is M, but the converse is not necessarily true.

☐ If M is orientable, then so is $\widetilde{M}$, but the converse is not necessarily true.

☐ The covering manifold $\widetilde{M}$ is orientable if and only if the base manifold M is orientable.

(9) Is every codimension-one submanifold M_0 of an orientable manifold M orientable?

☐ Yes, because then M_0 is the preimage of a regular value of a function $f : M \to \mathbb{R}$.

☐ Yes, because submanifolds of orientable manifolds are always orientable.

☐ No. The real projective plane $\mathbb{RP}^2$, as a submanifold of real projective space $\mathbb{RP}^3$, is a counterexample.

(10) Let $M_0 \subset M$ be a submanifold of codimension ≥ 2 and let $M \setminus M_0$ be oriented. Is M necessarily orientable?

☐ Yes. The charts (U, h) on M that are orientation-preserving on $U \setminus M_0$ form an orienting atlas on M.

☐ No. A counterexample is $\{p\} \subset \mathbb{RP}^2$.

☐ No. A counterexample is $\mathbb{RP}^2 \subset \mathbb{RP}^4$.

4.6 Exercises

EXERCISE 4.1. Let V be a real vector space with dimension $n \geq 1$, and let $(v_1, \ldots, v_{n-1}, v_n)$ and $(v_1, \ldots, v_{n-1}, v'_n)$ be two bases that differ only in their last vector. For $0 \leq t \leq 1$, we now set $v^t_n := (1-t)v_n + tv'_n$. Show that $(v_1, \ldots, v_{n-1}, v^t_n)$ is a basis for every $t \in [0, 1]$ if and only if $(v_1, \ldots, v_{n-1}, v_n)$ and $(v_1, \ldots, v_{n-1}, v'_n)$ have the same orientation.

EXERCISE 4.2. Show that a connected manifold has at most two orientations.

EXERCISE 4.3. Let M be a nonorientable n-dimensional manifold and let $\omega \in \Omega^n M$. Show that $\omega_p = 0$ for some $p \in M$.

EXERCISE 4.4. Let $\gamma : [0, 1] \to M$ be a continuous curve in an oriented n-dimensional manifold and let

$$v : [0, 1] \to \bigcup_{p \in M} \mathfrak{B}(T_p M)$$

be a continuous frame field along γ, that is, a continuous (relative to charts) correspondence that assigns a basis $v(t) = (v_1(t), \ldots, v_n(t))$ of $T_{\gamma(t)}M$ to each $t \in [0, 1]$. Show that if $v(0)$ is positively oriented, so is each $v(t)$ for $t > 0$. As an application of this lemma, prove that the projective plane $\mathbb{RP}^2$ is not orientable.

4.7 Hints for the Exercises

FOR EXERCISE 4.1. It would be a good idea to consider the determinant of the endomorphism that takes $(v_1, \ldots, v_n)$ to $(v_1, \ldots, v_{n-1}, v_n^t)$. Of course, this endomorphism can be written (for instance) as a matrix with respect to $(v_1, \ldots, v_n)$.

FOR EXERCISE 4.2. The standard connectedness argument should be applied here. If this is unfamiliar to you, may I recommend [J:*Top.*], the bottom of p. 14 and the top of p. 15? You should also observe that a transition map is orientation-preserving if and only if its Jacobian is positive.

FOR EXERCISE 4.3. As we know (see Section 3.3), an automorphism $f_p : T_pM \xrightarrow{\cong} T_pM$ acts on ω_p by multiplication by its determinant, so ω_p responds with values of the same sign to two bases of T_pM if and only if the bases have the same orientation. So how could you try to use an $\omega \in \Omega^n M$ with $\omega_p \neq 0$ for all $p \in M$ (hypothesis for an indirect proof) to orient M, contradicting the hypothesis of nonorientability? It's pretty easy to come up with this. The work of formulating it precisely lies in proving that the family of orientations of the tangent spaces defined in this way is locally coherent.

FOR EXERCISE 4.4.

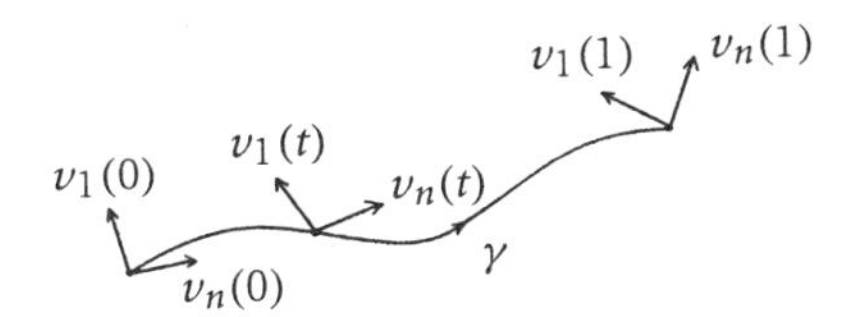

Figure 4.5.

The proof of the lemma about the continuous frame field along γ is, after Exercises 4.2 and 4.3, the third variation on the theme "the orientation can't suddenly switch." The real problem is the application to the orientability of the projective plane. Problems are often more transparent, in fact not infrequently easier to solve, if one generalizes them a bit. Here, for instance, it's useful to consider in what circumstances a quotient M/τ of a (path-connected) manifold under a fixed-point-free involution may be orientable and when it may not. (See Section 1.6.) All that remains to be proved in the concrete example is that the antipodal involution on S^2 is orientation-reversing. What about other dimensions?

5 CHAPTER Integration on Manifolds

5.1 What Are the Right Integrands?

Integration over n-dimensional manifolds reduces through charts to integration in $\mathbb{R}^n$. The objects integrated on oriented manifolds are n-forms, for the following reason. For an ordinary function $f : M \to \mathbb{R}$, the contribution of a chart domain U to the integral would clearly depend on the choice of chart h. But for an n-form, the integral of its component function pulled down by an orientation-preserving chart is independent of the coordinates, as we see from the change-of-variables formula for multiple integrals in $\mathbb{R}^n$. This is the main content of Chapter 5. Section 5.4 contains the technical details and Section 5.3 a summary of necessary background. In the first two sections we give an intuitive view of integration on manifolds.

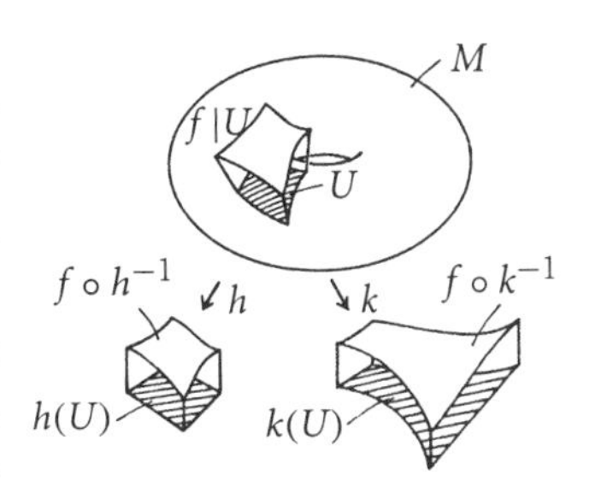

Figure 5.1. The integral of the downstairs function over the image of the chart domain obviously depends on the choice of chart.

Densities are natural candidates for the role of the integrand. Imagine a substance finely distributed throughout the manifold. Integrating the density of the distribution ought to give the total mass of the substance. What kind of mathematical object describes the density?

To approach this question we consider its infinitesimal, or linear-algebraic, version. Let V be an n-dimensional vector space (later T_pM), with a substance *uniformly* distributed throughout it. If we were dealing with $\mathbb{R}^n$, we could describe the density by the number measuring the amount of the substance in the unit cube $[0, 1]^n$. But in T_pM or V, instead of a distinguished unit cube we have only n-spans with equal rights.

Definition. Let $v_1, \ldots, v_k$ be elements of an n-dimensional real vector space V. Then

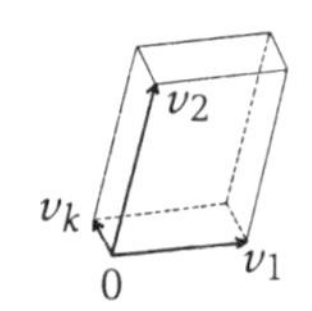

Figure 5.2. Span

$$\operatorname{span}(v_1, \ldots, v_k) := \left\{ \sum_{i=1}^{k} \lambda_i v_i : 0 \leq \lambda_i \leq 1 \right\}$$

is called the parallelepiped spanned by $v_1, \ldots, v_k$, their ***k-span***, or just their ***span***. The word "span" is generally used for the set of linear combinations of v_i with unrestricted λ_i. In that sense the v_i span an entire subspace of V. But we will use "span" as a noun only to mean a parallelepiped, and in context our usage should cause no confusion.

Without choosing a basis, we can describe the density, for example, by the map $\rho : V \times \cdots \times V \to \mathbb{R}$ that measures the amount of the substance contained in the span of any n vectors. What maps can arise in this way? In the attempt to formulate the notion of density mathematically, *positive homogeneity* and *shear invariance* are surely not too much to require.

Definition. Let V be an n-dimensional real vector space. We call a map $\rho : V^n = V \times \cdots \times V \to \mathbb{R}$ a ***density*** on V if it is ***positive-homogeneous*** and ***shear invariant***, i.e. if (1) $\rho(v_1, \ldots, \lambda v_i, \ldots, v_n) = |\lambda| \rho(v_1, \ldots, v_n)$ and (2) $\rho(v_1, \ldots, v_{i-1}, v_i + v_j, v_{i+1}, \ldots, v_n) = \rho(v_1, \ldots, v_n)$ for all $v_1, \ldots, v_n \in V$, $\lambda \in \mathbb{R}$, and $i \neq j$.

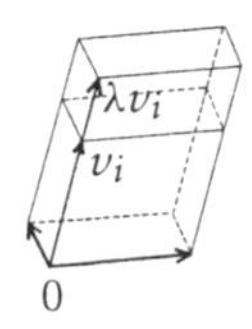

Figure 5.3a. Positive homogeneity

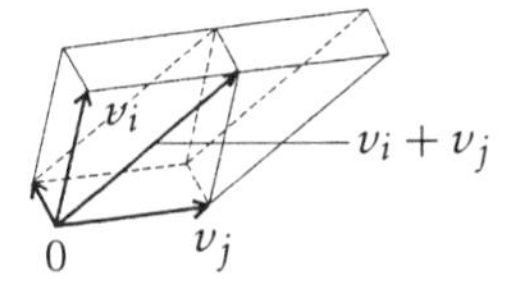

Figure 5.3b. Shear invariance

So far, such a density on V looks almost the same as an alternating n-form on V. The only difference is that transposing two vectors has no effect on the density (which depends only on the span) but reverses the sign of an n-form. The next lemma makes this more precise.

Lemma. *Let V be an n-dimensional vector space. If we choose an orientation $\mathcal{O}$ for V and change every map $\rho : V\times\cdots\times V \to \mathbb{R}$ to $\rho_{\mathcal{O}}$ by defining*

$$\rho_{\mathcal{O}} := \begin{cases} -\rho(v_1, \dots, v_n) & \textit{if } (v_1, \dots, v_n) \textit{ is negatively oriented,} \\ \rho(v_1, \dots, v_n) & \textit{otherwise,} \end{cases}$$

then ρ is a density if and only if $\rho_{\mathcal{O}}$ is an alternating n-form.

PROOF. "$\Leftarrow$" is trivial. To prove "$\Rightarrow$," let ρ be a density. Then (1) and (2) give

(3) $\rho(v_1, \dots, v_i + w, \dots, v_n) = \rho(v_1, \dots, v_n)$ if w is a linear combination of the variables $v_1, \dots, v_{i-1}, v_{i+1}, \dots, v_n$, and

(4) ρ is invariant under transposition of any two variables, hence under any permutation of the variables.

It also follows from (3) and (1) that ρ vanishes whenever $v_1, \dots, v_n$ are linearly dependent.

Now let $(e_1, \dots, e_n)$ be a positively oriented basis of V and let $\omega \in \text{Alt}^n V$ be the well-defined alternating n-form satisfying

$$\omega(e_1, \dots, e_n) = \rho(e_1, \dots, e_n).$$

We show by induction on k that

$$\omega(v_1, \dots, v_k, e_{k+1}, \dots, e_n) = \rho_{\mathcal{O}}(v_1, \dots, v_k, e_{k+1}, \dots, e_n)$$

for $k = 0, \dots, n$. For the induction step from k to $k+1$, assume without loss of generality that $(v_1, \dots, v_{k+1}, e_{k+2}, \dots, e_n)$ is linearly independent. We may assume by (3) that $v_1, \dots, v_{k+1}$ are in the linear hull V_{k+1} of $e_1, \dots, e_{k+1}$ and by (4) that $v_{k+1} \notin V_k$. Using (3) again, we see that $v_1, \dots, v_k$ are elements of V_k, and hence by dimension count that they span V_k. Another application of (3) therefore allows us to assume without loss of generality that $v_{k+1} = \lambda e_{k+1}$. In view of (1), this completes the induction step. □

Thus the space of densities on V, which we may as well call $\mathrm{Dens}(V)$, is, like $\mathrm{Alt}^n V$, a one-dimensional vector space. But there is no canonical isomorphism $\mathrm{Dens}(V) \cong \mathrm{Alt}^n(V)$ until one of the two orientations of V has been chosen.

We now make a definition analogous to that of n-forms on manifolds.

Definition. A ***density*** on an n-dimensional manifold is a correspondence ρ that assigns to every $p \in M$ a density

$$\rho_p \in \mathrm{Dens}(T_pM)$$

in the tangent space at p.

A density ρ on M is of course called *continuous*, *differentiable*, and so on if it is continuous or differentiable relative to charts, that is, if each $\rho(\partial_1, \dots, \partial_n)$ has the property. Because of its close affinity to $\Omega^n M$, we could denote the space of differentiable densities on M by $\Omega^{\mathrm{dens}} M$.

On *oriented* manifolds there is only a formal distinction between densities and n-forms, and the lemma above gives us a canonical bijection between $\Omega^{\mathrm{dens}} M$ and $\Omega^n M$. But passing to the opposite orientation reverses the sign of this bijection, so there seems to be an essential difference between densities and n-forms on *nonorientable manifolds*—and indeed there is.

Thus densities look like obvious integrands. Although n-forms do the same thing on oriented manifolds—and now we can see that this is why n-forms have something to do with integration—densities lead to a well-defined notion of integrals on nonorientable manifolds as well. But forms are preferred despite this, largely because they are also available as k-forms for $k < n$. Stokes's theorem, for instance, is a theorem about $(n-1)$-forms. Although k-densities can also be defined suitably so as to give integral theorems on nonorientable manifolds, the concept of forms would still be needed for the definition. (In a different language, let $L \to M$ denote the line bundle over the orientation double covering whose sections are what the physicists call pseudoscalars. Then densities are L-valued n-forms, and more generally k-densities would have to be interpreted as L-valued k-forms.)

5.2 The Idea behind the Integration Process

Although we could pursue the following reflections just as well for a density on an unoriented manifold, we stick to forms in view of what we intend to do later. So let M be an *oriented* n-dimensional manifold and ω a form on M. Each $\omega_p \in \mathrm{Alt}^n T_pM$ responds to oriented spans in T_pM, and we now try to understand whether and to what extent ω gives us a "response" $\int_M \omega$ to the whole manifold. To do this, we consider an orientation-preserving chart $h : U \xrightarrow{\cong} U' \subset \mathbb{R}^n$ on M and, in the image U' of the chart domain U, a rectangular parallelepiped, or box, $B' = [a^1, b^1] \times \cdots \times [a^n, b^n] \subset U'$. Subdividing the intervals $[a^i, b^i]$ determines a fine grid of many subboxes whose union is the large box B'. We call the preimages under the chart h the ***cells*** of the grid. To fix notation, we write σ_p for the cell with the "lower left vertex" p; this is the preimage of the subbox

$$\prod_{i=1}^{n} [x_p^i, x_p^i + \Delta x_p^i]$$

of the grid covering B', where $x_p^1, \ldots, x_p^n$ denote the coordinates of the lattice point $p \in B$. Of course it should be true

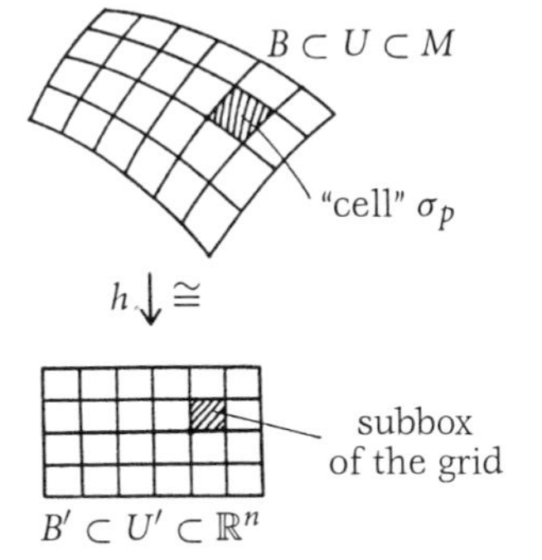

Figure 5.4. "Cells"

that

$$\int_B \omega = \sum_{p \in \text{lattice}} \int_{\sigma_p} \omega,$$

so next we try to understand whether and how ω responds to the individual cells.

If we want to follow the usual procedure of infinitesimal calculus, we should approximate the little cells linearly. We can do this by comparing each σ_p with the tangential span s_p in T_pM obtained as the preimage under dh_p (the linear approximation of the chart) of the subbox corresponding to σ_p. Since the unit vectors of $\mathbb{R}^n$ just correspond under the differential of the chart to the coordinate basis vectors $\partial_1, \ldots, \partial_n$ of the tangent space, the $\Delta x_p^\mu \cdot \partial_\mu$ are the edge vectors of the span. Thus

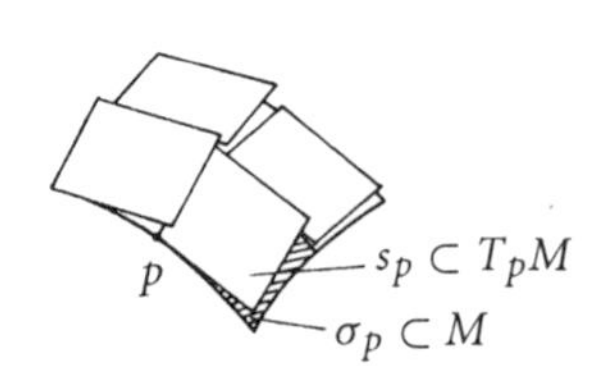

Figure 5.5. Approximating cells by tangential spans

$$s_p = \text{span}(\Delta x_p^1 \cdot \partial_1, \ldots, \Delta x_p^n \cdot \partial_n).$$

Now the alternating n-form ω_p on T_pM gives us a well-defined response

$$\omega_p(\Delta x_p^1 \cdot \partial_1, \ldots, \Delta x_p^n \cdot \partial_n) = \omega_p(\partial_1, \ldots, \partial_n)\Delta x_p^1 \cdot \ldots \cdot \Delta x_p^n,$$

and of course it seems natural to think of

$$\sum_{p \in \text{lattice}} \omega_p(\partial_1, \ldots, \partial_n)\Delta x_p^1 \cdot \ldots \cdot \Delta x_p^n$$

as an approximating sum for $\int_B \omega$ and to think of the integral as the limit of such sums as the grid covering B becomes finer and finer. Our statement at the beginning of this chapter about the reduction of integration over n-dimensional manifolds to integration in $\mathbb{R}^n$ translates into the formula

$$\int_B \omega = \int_{B'} (\omega_{1\ldots n} \circ h^{-1}) dx^1 \ldots dx^n,$$

which we can now understand geometrically.

Thus small cells are approximated by oriented tangential spans, and of course ω has a response ready for them—to a first approximation, ω responds to the oriented cells themselves. If we imagine the whole manifold as divided into small cells, then the integral is the sum of the responses to the cells, and we can trust that the result will be independent of the choice of charts used in the process because of our interpretation of the n-form as a density.

This picture of n-forms and the integral $\int_M \omega$ will turn out to be useful, especially for an intuitive understanding of the Cartan, or *exterior*, derivative and Stokes's theorem that $\int_M d\omega = \int_{\partial M} \omega$. But this doesn't mean that approximating cells by spans is technically the best path to follow for the actual introduction of the integral. In fact, we assume that integration theory in $\mathbb{R}^n$ is known and *exploit* it for integration on manifolds, rather than developing integration on manifolds analogously from scratch. The results we need from integration theory are listed in the next section.

5.3 Lebesgue Background Package

For the first time in quite a while, I'm putting additional demands on your background knowledge—this time by assuming some familiarity with the Lebesgue integral in $\mathbb{R}^n$. But I'm packaging the following background for you so I can say what I mean a bit more precisely.

The ***Lebesgue-measurable*** subsets of $\mathbb{R}^n$ form a ***σ-algebra*** $\mathfrak{M}$ on which the ***Lebesgue measure*** $\mu : \mathfrak{M} \to [0, \infty]$ is defined, thus turning $\mathbb{R}^n$ into a ***measure space*** $(\mathbb{R}^n, \mathfrak{M}, \mu)$. As in any measure space, the functions $\mathbb{R}^n \to \mathbb{R}$ that are ***integrable*** with respect to μ then form a vector space $\mathcal{L}^1(\mathbb{R}^n, \mu)$ on which the ***integral*** is given as a linear map, which we denote simply by

$$\begin{aligned} \mathcal{L}^1(\mathbb{R}^n, \mu) &\longrightarrow \mathbb{R}, \\ f &\longmapsto \int_{\mathbb{R}^n} f(x)dx. \end{aligned}$$

The map

$$\begin{aligned}\mathcal{L}^1(\mathbb{R}^n, \mu) &\longrightarrow \mathbb{R}_+, \\ f &\longmapsto \int_{\mathbb{R}^n} |f(x)|dx =: |f|_1\end{aligned}$$

is a seminorm on $\mathcal{L}^1$, and $|f|_1 = 0$ if and only if f vanishes ***almost everywhere***, i.e. except on a set of measure zero. Taking the quotient of $\mathcal{L}^1(\mathbb{R}^n, \mu)$ by the vector subspace of functions that vanish almost everywhere gives a normed vector space, which we denote by $L^1(\mathbb{R}^n, \mu)$, whose elements are the equivalence classes of integral functions under the relation of being equal almost everywhere.

Of course, much could be said about the properties of this Lebesgue integral: little lemmas and big theorems. I want at least to remind you of three wonderful convergence theorems, which incidentally hold for the Lebesgue integral on arbitrary measure spaces. These are the ***norm convergence theorem***, the ***monotone convergence theorem***, and the ***dominated convergence theorem*** (also called the ***Lebesgue convergence theorem***).

These three convergence theorems all deal with when a sequence of integrable functions converges to an integrable function and when a limit and an integral can be interchanged. By the *norm convergence theorem*, I mean the statement that $L^1(\mathbb{R}^n, \mu)$ is complete and hence a Banach space. The second theorem says that boundedness of the sequence of integrals $\int_{\mathbb{R}^n} f_k dx$ under monotone pointwise convergence $f_k \nearrow f$ implies the desired convergence statement. The third theorem guarantees that under arbitrary pointwise convergence $f_k \to f$, the existence of a dominating function $g \in \mathcal{L}^1$ (one for which $|f_k(x)| \le g(x)$ for all k and x) implies that $f \in \mathcal{L}^1$ and $\int f\, dx = \lim \int f_k dx$.

In addition to these three general convergence theorems, I want to remind you of two important theorems that pertain to $\mathbb{R}^n$ in particular, namely ***Fubini's theorem*** and the ***change-of-variables formula***. As you know, Fubini's theorem reduces integration on $\mathbb{R}^n$ inductively to integration on $\mathbb{R}$ ("iterated integrals"). I won't write down the exact wording of the theorem now. But the change-of-variables formula is crucial to integration on manifolds and should be cited in complete detail. First some terminology and notation.

Our discussion up to now has been about integrals on all of $\mathbb{R}^n$. But this includes the case of a subset $\Omega \subset \mathbb{R}^n$ as region of integration, in the following way: If Ω is contained in the domain of definition of f, we define $f_\Omega : \mathbb{R}^n \to \mathbb{R}$ by

$$f_\Omega(x) = \begin{cases} f(x) & \text{for } x \in \Omega, \\ 0 & \text{otherwise,} \end{cases}$$

no matter whether or how f was previously defined outside Ω. If f_Ω is in $\mathcal{L}^1(\mathbb{R}^n, \mu)$, we say that f is ***integrable on*** Ω (keep in mind that this is with respect to Lebesgue measure μ_n), and we write

$$\int_\Omega f(x)dx := \int_{\mathbb{R}^n} f_\Omega(x)dx.$$

Theorem (Change-of-variables formula). *Let $\Omega \subset \mathbb{R}^n$ be open and $f : \Omega \to \mathbb{R}$ integrable on Ω; let $\widetilde{\Omega} \subset \mathbb{R}^n$ be another open subset and $\varphi : \widetilde{\Omega} \xrightarrow{\cong} \Omega$ a C^1 diffeomorphism. Then $f \circ \varphi \cdot |\det J_\varphi|$ is also integrable on $\widetilde{\Omega}$ and*

$$\int_\Omega f\,dx = \int_{\widetilde{\Omega}} (f \circ \varphi) \cdot |\det J_\varphi| dx,$$

where $J_\varphi : \widetilde{\Omega} \to M(n \times n, \mathbb{R})$ denotes the Jacobian matrix of φ.

The diffeomorphism $\varphi : \widetilde{\Omega} \to \Omega$ can be viewed as a sort of "reparametrization." We shouldn't expect f and $f \circ \varphi$ to have the same integral; on the contrary, we need a correction factor. And we shouldn't be surprised that this factor is exactly the absolute value of the Jacobian: after all, the Jacobian matrix is the linear approximation of the diffeomorphism φ, so in passing from small boxes in $\widetilde{\Omega}$ to their images in Ω, the volume is approximated by multiplication by $|\det J_\varphi|$.

A careful proof takes some effort, as you probably remember. One thing that comes out of it is that diffeomorphisms between open sets in $\mathbb{R}^n$, and transition maps in particular, take measurable sets to measurable sets and sets of measure zero to sets of measure zero. This is about to be useful.

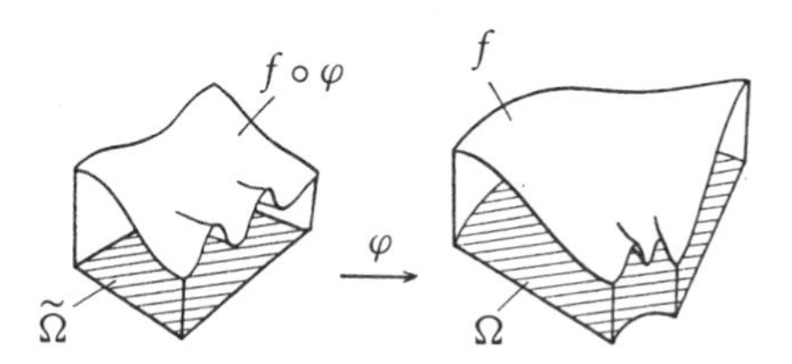

Figure 5.6. The change-of-variables formula

That's it for our Lebesgue package. If all its contents seem familiar to you, you're certainly well prepared for what follows. But I won't try to conceal that for the main goals of this course, Stokes's theorem and its consequences, you could get by with less integration theory: just the integral, Fubini's theorem, and the change-of-variables formula for C^∞ functions with compact support on $\mathbb{R}^n$ and the half-space. If you want to take this path, all you have to do now is work through Sections 9.5 and 9.6 instead of 5.4—don't worry, these sections are set up for this and *expect* a visit from Section 5.3—and you'll know enough about the notion of integrals on manifolds.

5.4 Definition of Integration on Manifolds

Definition. A subset A of an n-dimensional manifold is called ***measurable*** (resp. a ***set of measure zero***) if it has this property relative to charts, that is, if for some (hence every) covering of A by charts (U, h) on M, each $h(U \cap A)$ is Lebesgue-measurable (resp. a set of measure zero) in $\mathbb{R}^n$.

Thus the σ-algebra of Lebesgue-measurable sets is also well defined on a manifold, and its sets of measure zero are canonically recognizable. But note that of course we have no canonical measure on this σ-algebra.

Now let ω be an n-form on an oriented n-dimensional manifold M. To define $\int_M \omega$ we will decompose M into countably many small pieces, on each of which we can integrate by using a chart. These small pieces need not be coordinate cells, which would lead to major technical difficulties on the overlaps of adjacent charts. Instead, the good properties of the Lebesgue integral allow us great freedom in how we decompose M.

Terminology. For the following discussion, a subset $A \subset M$ will be called ***small*** if it is contained in a chart domain.

Note. *Any manifold can be decomposed into countably many pairwise disjoint small measurable subsets.* For instance, if $\mathfrak{A} = \{(U_i, h_i) : i \in \mathbb{N}\}$ is a countable atlas on M, then

$$A_1 := U_1,$$
$$A_{i+1} := U_{i+1} \setminus \bigcup_{k=1}^{i} A_k \quad \text{for } i \geq 1$$

gives such a partition $M = \bigcup_{i=1}^{\infty} A_i$.

Of course, we intend to set $\int_M \omega := \sum_{i=1}^{\infty} \int_{A_i} \omega$. As we have to integrate on small pieces by means of charts, it is already intuitively clear that the change-of-variables formula for the Lebesgue integral is what makes this technically feasible.

Theorem and Definition (Integration on manifolds). *An n-form ω on an oriented n-dimensional manifold M is called* ***integrable*** *if for some (hence every) decomposition $(A_i)_{i\in\mathbb{N}}$ of M into countably many small measurable subsets and some (hence every) sequence $(U_i, h_i)_{i\in\mathbb{N}}$ of orientation-preserving charts with $A_i \subset U_i$, the following holds: For every $i \in \mathbb{N}$, the downstairs* *component function*

$$a_i := \omega(\partial_1, \ldots, \partial_n) \circ h_i^{-1} : h_i(U_i) \to \mathbb{R}$$

of ω *relative to* (U_i, h_i) *is Lebesgue-integrable on* $h_i(A_i)$*, and*

$$\sum_{i=1}^{\infty} \int_{h_i(A_i)} |a_i(x)| dx < \infty.$$

The value

$$\sum_{i=1}^{\infty} \int_{h_i(A_i)} a_i(x) dx := \int_M \omega,$$

*which is independent of the decomposition and the charts, is called the **integral** of* ω *over* M*.*

PROOF OF THE ASSERTIONS. Let $(A_i)_{i\geq 1}$ and $(B_j)_{j\geq 1}$ be decompositions of M into measurable sets and let (U_i, h_i) and (V_j, k_j) be orientation-preserving charts with $A_i \subset U_i$ and $B_j \subset V_j$. Let ω be an n-form with downstairs component functions a_i (relative to (U_i, h_i)) and b_j (relative to (V_j, k_j)) such that ω satisfies the conditions in terms of the A_i's and h_i's: each a_i is integrable on $h_i(A_i)$ and $\sum_{i=1}^{\infty} \int_{h_i(A_i)} |a_i| dx < \infty$. We must show that each b_j is integrable on $k_j(B_j)$, that $\sum_{j=1}^{\infty} \int_{k_j(B_j)} |b_j| dx < \infty$, and that

$$\sum_{i=1}^{\infty} \int_{h_i(A_i)} a_i(x) dx = \sum_{j=1}^{\infty} \int_{k_j(B_j)} b_j(x) dx.$$

Recall that a Lebesgue-integrable function on $\mathbb{R}^n$ is also integrable on every measurable subset of $\mathbb{R}^n$. In particular, a_i is integrable on $h_i(A_i \cap B_j)$, and it follows from Lebesgue's convergence theorem that

$$\int_{h_i(A_i)} a_i \, dx = \sum_{j=1}^{\infty} \int_{h_i(A_i \cap B_j)} a_i \, dx,$$

and similarly for $|a_i|$ in place of a_i. Now, to pass from a_i on $h_i(A_i \cap B_j)$ to b_j on $k_j(A_i \cap B_j)$, we apply the change-of-variables formula cited in such detail in Section 5.3. That is, we set

$$\Omega := h_i(U_i \cap V_j),$$

$$f(x) := \begin{cases} a_i(x) & \text{for } x \in h_i(A_i \cap B_j), \\ 0 & \text{otherwise,} \end{cases}$$

$$\widetilde{\Omega} := k_j(U_i \cap V_j), \quad \text{and finally}$$

$$\varphi := h_i \circ k_j^{-1} | k_j(U_i \cap V_j), \quad \text{the transition map.}$$

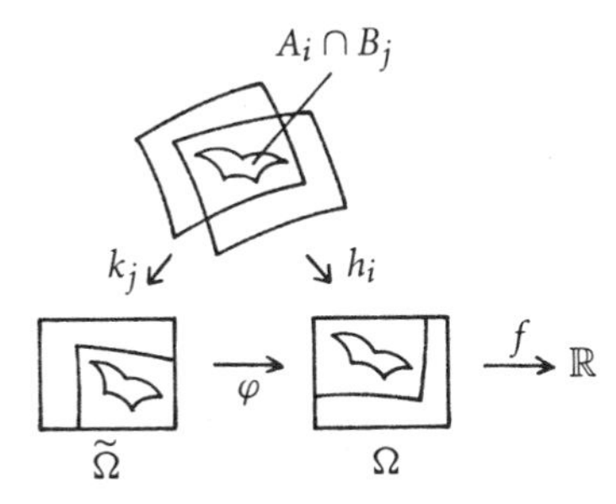

Figure 5.7.

For each $p \in U_i \cap V_j$, consider the three differentials

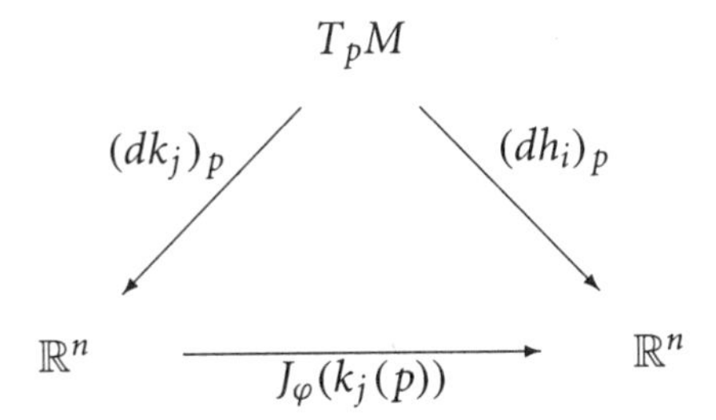

From the alternating n-form ω_p on T_pM, the (inverses of the) differentials of the two charts induce two alternating n-forms on $\mathbb{R}^n$, one of which takes the value $b_j(k_j(p))$ on the canonical basis, and the other the value $a_i(h_i(p))$. But the endomorphism $J_\varphi(k_j(p))$ acts on $\mathrm{Alt}^n\mathbb{R}^n$ by multiplication by the determinant, as we know from the lemma in Section 3.3, so we have

$$b_j(k_j(p)) = a_i(h_i(p)) \cdot \det J_\varphi(k_j(p)),$$

or

$$b_j = (a_i \circ \varphi) \cdot |\det J_\varphi|,$$

throughout $k_j(U_i \cap V_j)$. We can take the absolute value because φ preserves orientation and therefore has positive Jacobian. This also implies trivially that

$$(b_j)_{k_j(A_i \cap B_j)} = \left((a_i)_{h_i(A_i \cap B_j)} \circ \varphi\right) \cdot |\det J_\varphi|,$$

and similarly for $|a_i|$ and $|b_j|$ in place of a_i and b_j. So by the change-of-variables formula the function b_j is Lebesgue-integrable on $k_j(A_i \cap B_j)$ and

$$\int\limits_{k_j(A_i \cap B_j)} b_j \, dx = \int\limits_{h_i(A_i \cap B_j)} a_i \, dx,$$

and similarly for $|a_i|$ and $|b_j|$ in place of a_i and b_j. It follows that

$$\sum_{i,j=1}^{\infty} \int\limits_{k_j(A_i \cap B_j)} |b_j| dx < \infty.$$

In particular, $|b_j|$ and b_j are integrable even on $k_j(B_j)$ by the convergence theorems. We have

$$\int\limits_{k_j(B_j)} b_j \, dx = \sum_{i=1}^{\infty} \int\limits_{k_j(A_i \cap B_j)} b_j \, dx,$$

and similarly for $|b_j|$. Hence $\sum_{j=1}^{\infty} \int_{k_j(B_j)} |b_j| \, dx < \infty$, by the equation relating b_j and a_i that comes directly from the change-of-variables formula. It follows that

$$\sum_{j=1}^{\infty} \int\limits_{k_j(B_j)} b_j \, dx = \sum_{i=1}^{\infty} \int\limits_{h_i(A_i)} a_i \, dx,$$

as was to be proved. □

5.5 Some Properties of the Integral

How can you tell whether an n-form is integrable? We will usually be dealing with n-forms whose ***support***, the closed set

$$\operatorname{supp}\omega := \overline{\{p \in M : \omega_p \neq 0\}} \subset M,$$

is compact. If M itself is compact, for example, then of course *all* n-forms have compact support.

Lemma. *An n-form ω with compact support on an n-dimensional oriented manifold M is integrable if and only if it is* ***locally integrable****; that is, if around any point there is a chart (U, h) such that the downstairs component function*

$$\omega(\partial_1, \ldots, \partial_n) \circ h^{-1} : h(U) \longrightarrow \mathbb{R}$$

is Lebesgue-integrable on $h(U) \subset \mathbb{R}^n$.

PROOF. If $(U_i, h_i)_{i \in \mathbb{N}}$ is a countable atlas of such charts, then finitely many of them, say the first r, cover the support of ω. Set $A_1 := U_1$ and $A_{i+1} = U_{i+1} \setminus \bigcup_{k=1}^{i} A_k$. Then $\omega|A_i \equiv 0$ for all $i > r$, so

$$\sum_{i=1}^{\infty} \int_{h_i(A_i)} |a_i|\, dx = \sum_{i=1}^{r} \int_{h_i(A_i)} |a_i|\, dx < \infty. \qquad \square$$

Continuous n-forms are locally integrable, of course, and if ω is locally integrable and $A \subset M$ is measurable, then the form ω_A defined by

$$p \longmapsto \begin{cases} \omega_p, & p \in A, \\ 0 & \text{otherwise} \end{cases}$$

is also locally integrable. Thus the lemma already gives us a number of examples of integrable forms. In particular, on a compact oriented n-dimensional manifold M all continuous n-forms are integrable. A fortiori, so are all differentiable n-forms and hence all $\omega \in \Omega^n M$.

As we expect from the discussion in Section 5.1 and can now easily read off from the definition, reversing the orientation changes the sign of the integral:

Note. $\int\limits_{-M} \omega = -\int\limits_{M} \omega.$

What is the change-of-variables formula for integrals on manifolds? Instead of a diffeomorphism $\varphi : \widetilde{\Omega} \to \Omega$ between two open subsets of $\mathbb{R}^n$, we now consider an orientation-preserving diffeomorphism $\varphi : \widetilde{M} \xrightarrow{\cong} M$. If, for the integration on M, we use a decomposition $M = \cup A_i$ and charts (U_i, h_i) around the A_i's, then for $\widetilde{M}$ we can use the corresponding data under φ, namely the decomposition $\widetilde{M} = \cup\varphi^{-1}(A_i)$ and the charts $(\varphi^{-1}(U_i), h_i \circ \varphi|\varphi^{-1}(U_i))$. Then the n-forms ω on M and $\varphi^*\omega$ on $\widetilde{M}$ have exactly the same downstairs component functions, and the nice and important naturality property of the integral follows.

Note ("Change-of-variables formula" for integration on manifolds). *If $\varphi : \widetilde{M} \xrightarrow{\cong} M$ is an orientation-preserving diffeomorphism between oriented n-dimensional manifolds, then an n-form ω on M is integrable if and only if $\varphi^*\omega$ is integrable on $\widetilde{M}$, and we have*

$$\int\limits_{M} \omega = \int\limits_{\widetilde{M}} \varphi^*\omega.$$

Of course, this was also to be expected from the intuitive discussion of the integral in Section 5.2, since the induced form $\varphi^*\omega$, by definition, responds to a cell (span) as ω does to the image cell.

Finally, as far as integrability and the integral on subsets A of M are concerned, we follow the spirit of the convention we established for the Lebesgue integral when we recalled the change-of-variables formula in Section 5.3. We define $\int_A \omega := \int_M \omega_A$, where ω_A agrees with ω on A and is set equal to zero outside A. The change-of-variables formula now takes the following form.

Corollary. *If $\varphi : \widetilde{M} \to M$ is an orientation-preserving diffeomorphism and $A \subset \widetilde{M}$, then ω is integrable on $\varphi(A)$ if and only if $\varphi^*\omega$ is integrable on A, and we have*

$$\int_A \varphi^*\omega = \int_{\varphi(A)} \omega.$$

One last general remark. We used charts here to reduce integration on manifolds to integration on $\mathbb{R}^n$, so we had to assume only the latter as known. But whoever knows the Lebesgue integral for arbitrary measure spaces holds a master key that gives direct access to the general properties of the integral on manifolds—the convergence theorems, for instance.

What I mean is this. On any oriented manifold one can construct a *volume form*, a nowhere-vanishing n-form $\omega_M \in \Omega^n M$ that gives a positive response to positively oriented bases. This is quite easy by means of a *partition of unity*, which we will encounter as a tool in connection with Stokes's theorem. Let ω_M be any such volume form. Then through

$$\mu(X) := \int_X \omega_M$$

it defines a measure μ on the σ-algebra $\mathfrak{M}$ of Lebesgue-measurable subsets of the manifold M and turns M into a measure space. A function $f : M \to \mathbb{R}$ is integrable on this measure space if and only if the n-form $f\omega_M$ is integrable, and then we have

$$\int_M f \, d\mu = \int_M f \, \omega_M.$$

But since $\dim \mathrm{Alt}^n T_p M = 1$, *every* n-form on M is $f\omega_M$ for some f, so integrating n-forms on oriented manifolds can also be viewed as integrating functions on a measure space. A volume form is not, however, given canonically.

5.6 Test

(1) The span of the three unit vectors in $\mathbb{R}^3$ is a

☐ tetrahedron. ☐ triangle. ☐ cube.

(2) If $A : \mathbb{R}^n \to \mathbb{R}^n$ is a linear map, then the n-dimensional volume of $A([0, 1]^n)$ is

☐ $\|A\|$.

☐ $|\det A|$.

☐ $|a_{11} \cdot \ldots \cdot a_{nn}|$.

(3) If ρ is a density and ω is an alternating n-form on an n-dimensional vector space V, then

☐ $-|\rho|$ is an alternating n-form.

☐ $-|\omega|$ is a density.

☐ $|\omega|$ is a density.

(4) A set $X \subset \mathbb{R}^n$ has measure zero if and only if for every ε there is a sequence of cubes W_i, with total volume $\sum_{i=1}^{\infty} \mathrm{Vol}(W_i) \leq \varepsilon$, such that

☐ $X \subset \cup_{i=1}^{\infty} W_i$.

☐ $X \subset \cap_{i=1}^{\infty} W_i$.

☐ $X \subset W_i$ for sufficiently large i.

(5) In the plane $\mathbb{R}^2$, let R denote the rectangle $(1, 2) \times (0, \frac{\pi}{2})$ and K the quarter-annulus in the first quadrant with radii 1 and 2. Then the change from polar to Cartesian coordinates, $(r, \varphi) \mapsto (x, y)$, given by $x = r \cos \varphi$ and $y = r \sin \varphi$, defines a diffeomorphism Φ from

☐ K to itself. ☐ K to R. ☐ R to K.

(6) For Φ as in (5), the Jacobian $J_{\Phi}(r, \varphi)$ is

☐ $r \sin 2\varphi$. ☐ r. ☐ $-r$.

(7) You will occasionally come across the notational custom of denoting a certain function in the most disparate coordinate systems over and over again by f, as though the convention

$$f(x_1', \dots, x_n') = f(x_1'(x_1, \dots, x_n), \dots, x_n'(x_1, \dots, x_n)) \overset{!}{=}: f(x_1, \dots, x_n)$$

had been accepted. Very confusing! But with a little effort you can figure out what it means. Imagine that f actually lives, independently of coordinates, on U (for instance, on a region U of real physical space), that $f(x_1', \dots, x_n')$ means the value of the function at the point that has coordinates $(x_1', \dots, x_n')$ *with respect to the primed coordinate system*, and so on. Then $f \circ h^{-1}$, $f \circ h'^{-1}$, etc., are always just written f, suppressing the names h, h', ... of the charts.

We don't really want to adopt this notation, but we do want to be able to read it if necessary, and in this sense the question now asks: How does the change-of-variables formula for integrals read between Cartesian and polar coordinates if the convention above is applied?

☐ $\iint f(x, y) dx dy = \iint f(r, \varphi) r \, dr \, d\varphi$.

☐ $\iint f(x, y)\sqrt{x^2 + y^2} dx dy = \iint f(r, \varphi) dr \, d\varphi$.

☐ $\iint f(x, y) dx dy = \iint f(r, \varphi) dr \, d\varphi$.

(8) In the local coordinates of a chart (U, h), the integral of an n-form ω over the chart domain is

$$\int_U \omega = \int_{h(U)} f(x) dx,$$

where $f : h(U) \to \mathbb{R}$ can be given by

☐ $f(x) = \omega(h^{-1}(x))$.

☐ $f(x^1, \dots, x^n) = \omega_{1 \dots n}$ (Ricci calculus).

☐ $f \circ h = \omega(\partial_1, \dots, \partial_n)$.

(9) If two charts (U, h) and (U, h') differ only in the sign of their first coordinate and if a and a' are the downstairs component functions of an n-form ω given on U, then

$$\int_{h(U)} a\,dx = -\int_{h'(U)} a'\,dx.$$

Why?

- ☐ Because
$$a(x^1, \ldots, x^n) = a'(-x^1, x^2, \ldots, x^n)$$
in the coordinates of $\mathbb{R}^n$, and the Jacobian of the transition map is -1.
- ☐ Because $a(x^1, \ldots, x^n) = -a'(-x^1, x^2, \ldots, x^n)$, and the absolute value of the Jacobian of the transition map is 1.
- ☐ Because $a(x^1, \ldots, x^n) = -a'(x^1, x^2, \ldots, x^n)$, and the transition map is orthogonal.

(10) For orientation-reversing diffeomorphisms $\varphi : M \to N$, we have

- ☐ $\int_M \omega + \int_{\varphi(M)} \varphi^*\omega = 0.$
- ☐ $\int_M \varphi^*\omega + \int_{\varphi(M)} \omega = 0.$
- ☐ $\int_M \omega + \int_{\varphi^{-1}(N)} \varphi^*\omega = 0.$

5.7 Exercises

EXERCISE 5.1. Give an n-form ω on $\mathbb{R}^n$ such that $\int_A \omega = \mu(A)$ for every $A \subset \mathbb{R}^n$ with Lebesgue measure $\mu(A) < \infty$.

EXERCISE 5.2. Let ω be an integrable n-form on the oriented n-dimensional manifold M. Show that, just as for integration in $\mathbb{R}^n$, the following holds: If an n-form η agrees with ω almost everywhere on M, then η is also integrable, and $\int_M \omega = \int_M \eta$.

EXERCISE 5.3. Let M be an oriented n-dimensional manifold. As an analogue to $|\cdot|_1$ on $\mathcal{L}^1(\mathbb{R}^n, \mu)$, how would a seminorm

$|\cdot|_1$ have to be defined on the vector space $\mathcal{L}^1(M)$ of integrable n-forms on M? For each $\omega \in \mathcal{L}^1(M)$, define an n-form $|\omega|$ such that $|\omega|_1 := \int_M |\omega|$ determines a seminorm that is zero exactly on the forms that vanish almost everywhere.

EXERCISE 5.4. Let $\pi : \widetilde{M} \to M$ be an m-sheeted covering of the connected n-dimensional oriented manifold M. Let the covering manifold $\widetilde{M}$ be oriented in such a way that π is orientation-preserving everywhere. Show that if ω is integrable on M, then $\pi^*\omega$ is integrable on $\widetilde{M}$ and $\int_{\widetilde{M}} \pi^*\omega = m \int_M \omega$.

5.8 Hints for the Exercises

FOR EXERCISE 5.1. For the solution of this exercise, all you have to know about Lebesgue measure is that

$$\mu(A) = \int_A 1\, dx$$

for Lebesgue-measurable $A \subset \mathbb{R}^n$ with finite measure. Of course, this is assumed to be known and need not be proved here. The exercise isn't hard and is just intended to make you read through the definition of the integral of an n-form again.

FOR EXERCISE 5.2. The point of this exercise is the same. It's just that you can't get by here, as you could in Exercise 5.1, with a single chart on M.

FOR EXERCISE 5.3. Warning: Here $|\omega|_p$ does not mean the absolute value $|\omega_p|$ of $\omega_p : T_pM \times \cdots \times T_pM \to \mathbb{R}$. That wouldn't be an alternating n-form on T_pM. For every $p \in M$, though, you should set $|\omega|_p := \pm\omega_p$ with the right choice of sign; the only question is how the sign depends on p. Of course, you should prove that $|\omega|$ really is integrable for $\omega \in \mathcal{L}^1(M)$ and that $|\cdot|_1 := \int_M |\cdot|$ is a seminorm on $\mathcal{L}^1(M)$ with the stated property. What this means here is a reduction to the corresponding properties of the Lebesgue integral on $\mathbb{R}^n$.

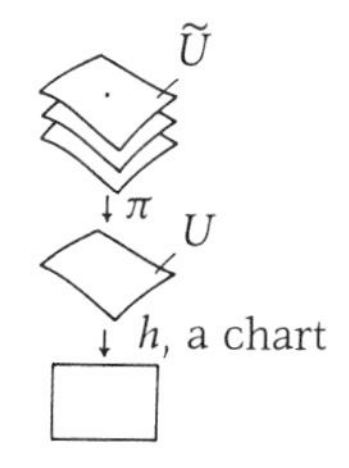

Figure 5.8. The differentiable structure on $\widetilde{M}$

FOR EXERCISE 5.4. You'll know enough about the notion of coverings for this exercise if you read pp. 126–131 in [J:*Top*] and notice in addition that in the case of a covering $\pi : \widetilde{M} \to M$ of a manifold M, the covering space $\widetilde{M}$ is also a manifold in a canonical way. In fact, its differentiable structure is the only one for which π is a local diffeomorphism everywhere. How will the decomposition

$$\widetilde{M} = \bigcup_{i=1}^{\infty} \bigcup_{j=1}^{m} \widetilde{A}_{ij}$$

of $\widetilde{M}$ into measurable subsets have to be chosen so that the integrability of $\pi^*\omega$ and the formula $\int_{\widetilde{M}} \pi^*\omega = m \int_M \omega$ follow easily? It's intuitively clear!

6 CHAPTER Manifolds-with-Boundary

6.1 Introduction

The classical version of Stokes's theorem deals with the connection between "surface integrals" and "line integrals." A three-dimensional version, called Gauss's integral theorem, makes a statement about the relationship between "volume integrals" and surface integrals.

Figure 6.1: In the original version $\int_M \operatorname{curl} \vec{v} \cdot d\vec{S} = \int_{\partial M} \vec{v} \cdot d\vec{s}$ of Stokes's theorem, the integration is over a surface and its boundary curve.

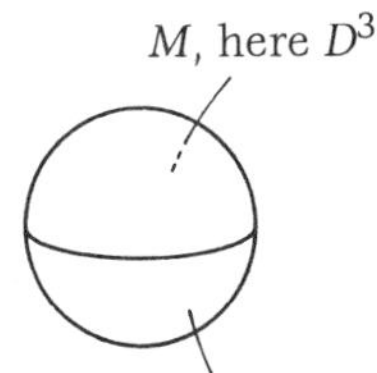

Figure 6.2: In Gauss's integral theorem $\int_M \vec{v} \cdot d\vec{S} = \int_M \operatorname{div} \vec{v}\, dV$, the integration is "over a closed surface and the volume it encloses."

Here, of course, we would like to treat both cases simultaneously, and even for this an n-dimensional version of the theorem would be worthwhile. Nor do we want to restrict ourselves to submanifolds of $\mathbb{R}^3$ or $\mathbb{R}^N$. In order to formulate Stokes's theorem in its full generality, we need the notion of *manifolds-with-boundary*, to which the present chapter is devoted.

6.2 Differentiability in the Half-Space

The local model for manifolds-with-boundary is the closed half-space, just as $\mathbb{R}^n$ is the local model for manifolds. To turn this notion into a precise definition, we must first explain what differentiability means in the case of the half-space.

Which half-space we use is unimportant, but in view of a certain orientation convention that will come up later, we choose the *left* half-space.

Notation and Terminology. For $n \geq 1$ we let $\mathbb{R}^n_-$ denote the half-space $\{x \in \mathbb{R}^n : x^1 \leq 0\}$ and $\partial\mathbb{R}^n_- := \{x \in \mathbb{R}^n : x^1 = 0\} = \{0\} \times \mathbb{R}^{n-1}$ its ***boundary***. If $U \subset \mathbb{R}^n_-$ is open in the subspace topology of $\mathbb{R}^n_- \subset \mathbb{R}^n$ (for short: *open in* $\mathbb{R}^n_-$ or *an open subset of* $\mathbb{R}^n_-$), then $\partial U := U \cap \partial\mathbb{R}^n_-$ is called the ***boundary*** of U and the elements $p \in \partial U$ the ***boundary points*** of U.

Of course, the boundary ∂U of U may be empty, and this obviously happens when $U \subset \mathbb{R}^n_-$ is open not only in $\mathbb{R}^n_-$ but also in $\mathbb{R}^n$.

Figure 6.3a. $\partial U \neq \emptyset$

Figure 6.3b. $\partial U = \emptyset$

In topology, what is meant by a *boundary point* of a subset A of a topological space X is an element $x \in X$ that is neither an interior nor an exterior point of X. But we should avoid this terminology for a while, because it clashes with the notion of boundary introduced above for an open subset U of $\mathbb{R}^n_-$. Observe that in general ∂U does *not* coincide with the topological boundary of U, no matter whether U is regarded as a subset of $\mathbb{R}^n_-$ or of $\mathbb{R}^n$.

Definition. Let U be open in $\mathbb{R}^n_-$. A map $f : U \to \mathbb{R}^k$ is called ***differentiable at the point*** $p \in U$ if it can be extended to a map that is differentiable in a neighborhood of p in $\mathbb{R}^n$; that is, if there exist an open neighborhood $\widetilde{U}_p$ of p in $\mathbb{R}^n$ and a differentiable map $g : \widetilde{U}_p \to \mathbb{R}^k$ such that $f|U \cap \widetilde{U}_p = g|U \cap \widetilde{U}_p$.

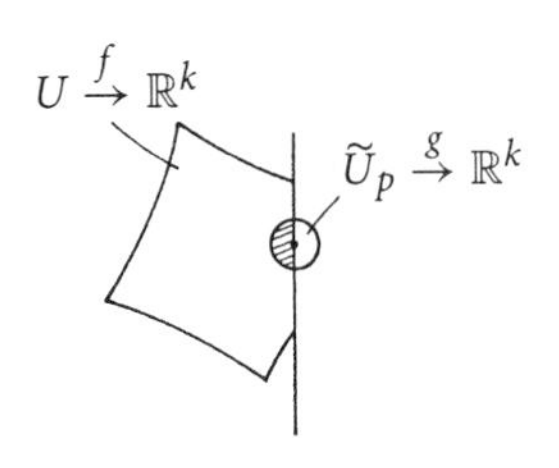

Figure 6.4. Differentiability at boundary points

For $p \in U \setminus \partial U$ this is nothing new, and f is simply ***differentiable***, i.e. differentiable everywhere, if it is differentiable in $U \setminus \partial U$ in the usual sense and differentiable for $p \in \partial U$ in the sense above. Of course, by a ***diffeomorphism*** between open subsets of $\mathbb{R}^n_-$ we mean a bijection that is differentiable in both directions. Such diffeomorphisms will be the transition maps of the as yet undefined manifolds-with-boundary. The following two lemmas clarify their boundary behavior.

6.3 The Boundary Behavior of Diffeomorphisms

Lemma 1. *If $f : U \xrightarrow{\cong} V$ is a diffeomorphism between open subsets of $\mathbb{R}^n_-$, then $f(\partial U) = \partial V$. Hence*

$$f|\partial U : \partial U \xrightarrow{\cong} \partial V$$

is a diffeomorphism between open subsets of $\mathbb{R}^{n-1}$.

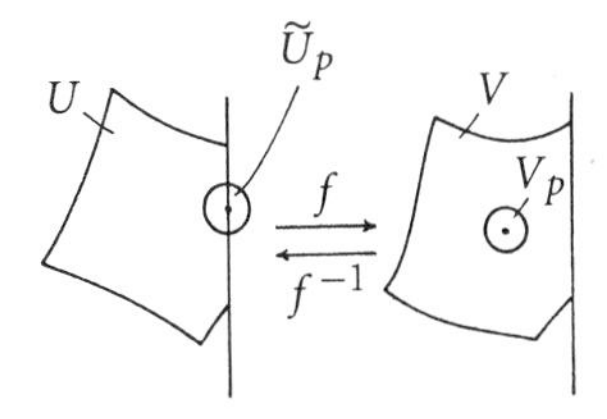

Figure 6.5. Assumption

PROOF. Let $p \in \partial U$ and let $g : \widetilde{U}_p \to \mathbb{R}^n$ be a local differentiable extension of f. Suppose $f(p)$ were not a boundary point of V. Then by the continuity of f^{-1} it would have a neighborhood V_p in V, open in $\mathbb{R}^n$, such that $f^{-1}(V_p) \subset \widetilde{U}_p$. But $g \circ (f^{-1}|V_p)$ is the identity on V_p, and g and $f^{-1}|V_p$ are differentiable in the usual sense. Hence f^{-1} has full rank n

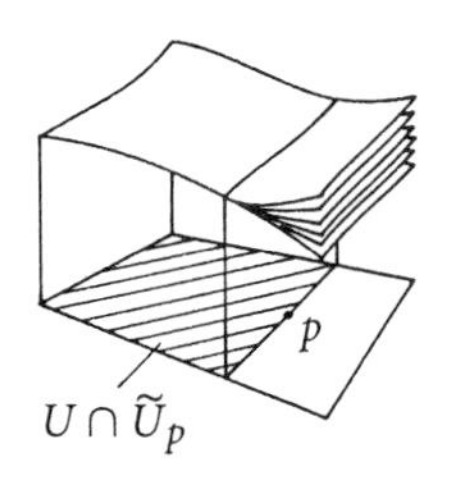

Figure 6.6. Local extension is not uniquely determined, but the $\partial_\alpha f|_p$ are.

at $f(p)$, so by the inverse function theorem it is a local diffeomorphism in the usual sense. In particular, $f^{-1}(V_p) \subset U$ is a neighborhood of p in $\mathbb{R}^n$, contradicting $p \in \partial U$. This shows that $f(\partial U) \subset \partial V$. But f is a diffeomorphism, and a similar argument applied to f^{-1} shows that $f^{-1}(\partial V) \subset \partial U$. Therefore $f(\partial U) = \partial V$. □

Of course, the local differentiable extension of a map $f : U \to \mathbb{R}^k$ around a boundary point p is not uniquely determined, but all the partial derivatives $\partial_\alpha f$ of f at the point p (and in particular the Jacobian $J_f(p)$) are.

Lemma 2. *If $f : U \xrightarrow{\cong} V$ is a diffeomorphism between open subsets of $\mathbb{R}^n_-$ and if $p \in \partial U$, then the well-defined differential*

$$df_p : \mathbb{R}^n \xrightarrow{\cong} \mathbb{R}^n$$

maps the subspace $\{0\} \times \mathbb{R}^{n-1}$ and each half-space $\mathbb{R}^n_\pm$ into itself; that is, the Jacobian matrix is of the form

$$\left(\begin{array}{c|c} \partial_1 f^1 & 0 \\ \hline \partial_1 f^2 & \\ \vdots & J_{f|\{0\}\times\mathbb{R}^{n-1}}(p) \\ \partial_1 f^n & \end{array}\right)$$

where $\partial_1 f^1 > 0$.

Proof. Since $f(\partial U) = \partial V$, we have $f^1|\partial U \equiv 0$, so $\partial_k f^1 = 0$ for $k = 2, \ldots, n$. Since V lies in $\mathbb{R}^n_-$, we have $f^1 \leq 0$ on U and thus

$$\frac{f^1(p + te_1) - f^1(p)}{t} \geq 0$$

for $t < 0$. Hence $\partial_1 f^1 \geq 0$, and in fact $\partial_1 f^1 > 0$ because $J_f(p)$ has full rank. □

6.4 The Concept of Manifolds-with-Boundary

Now we turn from future changes of charts to manifolds-with-boundary themselves. The only formal difference from ordinary manifolds ("without boundary") is that now we also admit open subsets of $\mathbb{R}^n_-$ as images under charts.

Let X be a topological space. A homeomorphism h from an open subset U of X onto an open subset U' of $\mathbb{R}^n_-$ or $\mathbb{R}^n$ will be called an ***n-dimensional chart-with-boundary*** for X. The notions of ***n-dimensional atlas-with-boundary, differentiable n-dimensional atlas-with-boundary,*** and ***n-dimensional differentiable structure-with-boundary*** (maximal atlas) are to be understood accordingly.

Definition. Let $n \geq 1$. An ***n-dimensional manifold-with-boundary*** is a pair $(M, \mathcal{D})$, usually abbreviated M, consisting of a second-countable Hausdorff space M and an n-dimensional differentiable structure-with-boundary $\mathcal{D}$ for M. We call maps between manifolds-with-boundary differentiable if they are differentiable relative to charts.

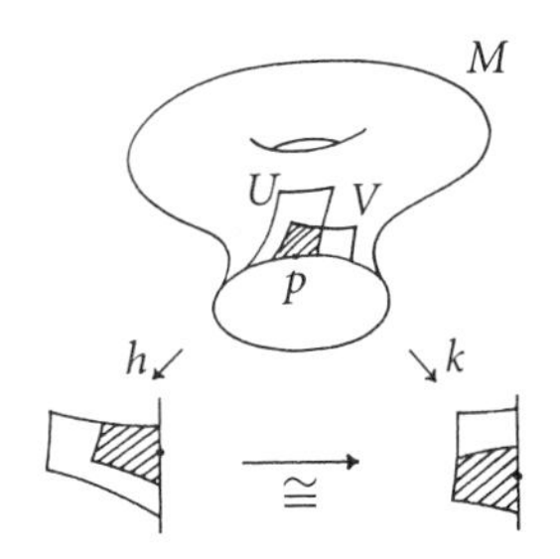

Figure 6.7. If p is a boundary point in terms of h, it is also a boundary point in terms of k: the boundary of M is well defined.

Transition maps must take boundary points to boundary points, as we saw in Lemma 1. This justifies the next definition.

Definition. Let M be a manifold-with-boundary. A point $p \in M$ is called a ***boundary point*** of M if it is mapped by some (hence every) chart (U, h) around p to a boundary point $h(p)$ of $h(U) \subset \mathbb{R}^n_-$. The set ∂M of boundary points is called the ***boundary*** of the manifold-with-boundary M.

Note. *If M is an n-dimensional manifold-with-boundary, the restrictions*

$$h|U \cap \partial M : U \cap \partial M \xrightarrow{\cong} \partial(h(U)) \subset \{0\} \times \mathbb{R}^{n-1} \cong \mathbb{R}^{n-1}$$

of the charts on M provide its boundary ∂M with an $(n-1)$-dimensional ordinary differentiable atlas and thus turn ∂M into an ordinary $(n-1)$-dimensional manifold (without boundary).

This manifold is what is always meant later on when the boundary ∂M of a manifold-with-boundary is mentioned. We also say that M is ***bounded by*** ∂M or that ∂M ***bounds*** M. If $f : M \to N$ is a differentiable map between manifolds-with-boundary, then of course $f|\partial M : \partial M \to N$ is also differentiable. Our next observation follows from Lemma 1.

Note. *If $f : M \xrightarrow{\cong} N$ is a diffeomorphism between manifolds-with-boundary, then $f(\partial M) = \partial N$ and $f|\partial M : \partial M \xrightarrow{\cong} \partial N$ is a diffeomorphism.*

For $n \geq 1$, any ordinary n-dimensional manifold M will be considered in the obvious way as a manifold-with-boundary with $\partial M = \emptyset$. By a ***zero-dimensional manifold-with-boundary*** we mean just an ordinary zero-dimensional manifold. Thus the boundary of a zero-dimensional manifold-with-boundary is empty—as a (-1)-dimensional manifold should be.

6.5 Submanifolds

We won't write out in detail all that can immediately be generalized from ordinary manifolds to manifolds-with-boundary. If that were necessary, we would have been better off basing everything from the beginning on the more general concept! There are things, however, whose generalization to manifolds-with-boundary involves certain decisions or conventions, or that for some other reason is not completely self-explanatory, and a few matters of this kind will be discussed in this and the following sections.

Definition. Let M be an n-dimensional manifold-with-boundary and let $1 \leq k \leq n$. A subset M_0 of M is called a k-dimensional submanifold-with-boundary if around any $p \in M_0$ there exists a chart-with-boundary (U, h) on M such that

$$h(U \cap M_0) = (\mathbb{R}^k_- \times \{0\}) \cap h(U).$$

This is not the only plausible way to interpret the notion of submanifolds for manifolds-with-boundary. In choosing this version, we make two decisions. First, we do not require that $\partial M_0 \subset \partial M$. But second, if a point $p \in M_0$ lies in the boundary of M, then it is also a boundary point of M_0, and M_0 is "transversal" to ∂M there, in the sense that relative to charts, M_0 and ∂M must abut at p as do $\mathbb{R}^k_-$ and $\{0\} \times \mathbb{R}^{n-1}$:

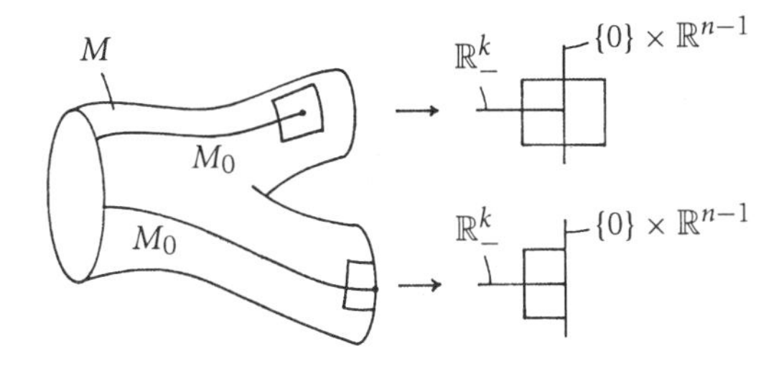

Figure 6.8. The two admissible possibilities for the position of ∂M_0 relative to ∂M

In particular, ∂M itself is *not* a submanifold of M unless it is empty, nor do we allow nonempty subsets of ∂M as submanifolds of M. To extend these decisions to $k = 0$, we require zero-dimensional submanifolds of M not to intersect the boundary. So a zero-dimensional submanifold M_0 of M will just be a zero-dimensional submanifold M_0 of $M \setminus \partial M$ in the ordinary sense.

Just as submanifolds of ordinary manifolds are themselves manifolds, k-dimensional submanifolds-with-boundary are k-dimensional manifolds-with-boundary in a canonical way. The restrictions of the flatteners (U, h) to each $U \cap M_0$ form a k-dimensional differentiable atlas on M.

6.6 Construction of Manifolds-with-Boundary

As examples of constructions of ordinary manifolds, we introduced sums, products, certain quotients, and the preim-

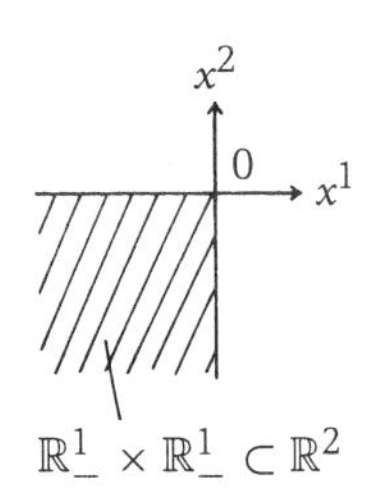

Figure 6.9. Edges can occur in the product.

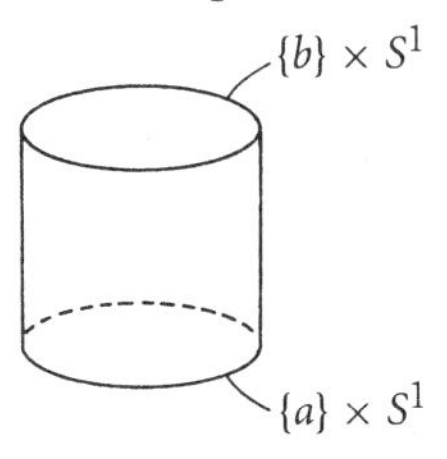

Figure 6.10. $[a, b] \times D^2$ with its "edges"

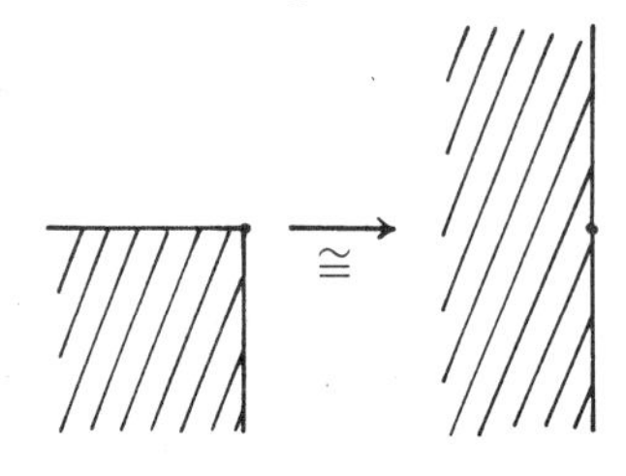

Figure 6.11. A smoothing map

ages of regular values. The disjoint sum $M_1 + M_2$ of two n-dimensional manifolds-with-boundary is itself an n-dimensional manifold-with-boundary, in a canonical way. There is a slight technical difficulty in taking products: although $\mathbb{R}^k \times \mathbb{R}^n = \mathbb{R}^{k+n}$ canonically, $\mathbb{R}^k_- \times \mathbb{R}^n_-$ is not a half-space but a *quadrant* in $\mathbb{R}^{k+n}$. For example, taking the product $[a, b] \times D^2$ of a closed interval and a closed disk gives a three-dimensional solid cylinder with two "edges" in its boundary, namely $\{a\} \times S^1$ and $\{b\} \times S^1$. More generally, $M \times N$ is intuitively something like a manifold that has boundary

$$\partial(M \times N) = \partial M \times N \cup M \times \partial N$$

and an "edge" (or "corner") along $\partial M \times \partial N$. *Why* one wants to consider such products in the first place determines whether one "smooths" the edges and turns them into genuine manifolds-with-boundary by using a homeomorphism $\mathbb{R}^1_- \times \mathbb{R}^1_- \to \mathbb{R}^2_-$ that fails to be a local diffeomorphism only at 0, or leaves them as they are and develops a theory of "manifolds-with-edges" (or "manifolds-with-corners"). We take neither of these paths here, but only point out that when at least one of the two factors has no boundary, the product is again a manifold-with-boundary in a canonical way.

The quotient M/τ of an n-dimensional manifold-with-boundary M under a fixed-point-free involution τ is canonically an n-dimensional manifold-with-boundary, by the same argument that was given in Section 1.6 for ordinary manifolds. Here $\partial(M/\tau) = (\partial M)/\tau$.

Just as for ordinary manifolds, the regular value theorem is an important source of concrete examples of manifolds-with-boundary.

Lemma. *If M is an n-dimensional manifold without boundary and $c \in \mathbb{R}$ is a regular value of a C^∞ function $f : M \to \mathbb{R}$, then $M_0 := \{p \in M : f(p) \leq c\}$ is an n-dimensional submanifold-with-boundary of M.*

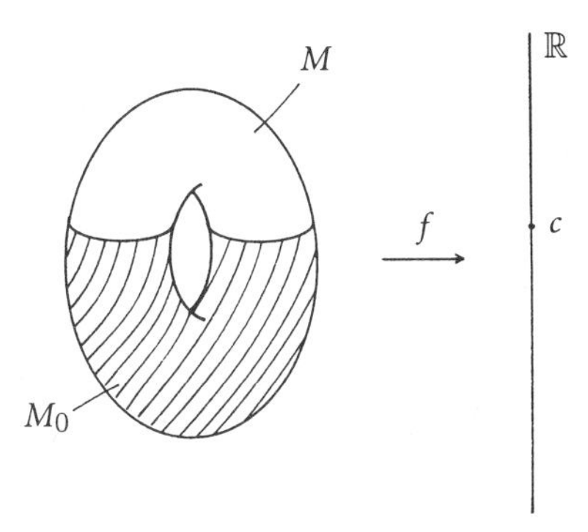

Figure 6.12. The preimage $f^{-1}((-\infty, c]) =: M_0$ at a regular value c. Here $f^{-1}(c) = \partial M_0$.

6.7 Tangent Spaces to the Boundary

What about the tangent spaces T_pM for boundary points $p \in \partial M$? Are they well defined at all? And if so, might it be better to use tangent half-spaces instead?

Remark and Convention. The tangent space for manifolds-with-boundary and at boundary points $p \in \partial M$ is also well defined by

$$T_pM := T_p^{\text{alg}}M \underset{\text{canon}}{\cong} T_p^{\text{phys}}M,$$

as in Chapter 2, and relative to a chart (U, h) the coordinate basis vectors $(\partial_1, \dots, \partial_n)$ of T_pM are defined for every $p \in U$. Thus we use the whole vector space T_pM as the tangent space at boundary points as well, but the two half-spaces

$$T_p^{\pm}M := (dh_p)^{-1}(\mathbb{R}^n_{\pm})$$

are well defined independently of the chart.

Note and Terminology. Let p be a boundary point of M. Then it is clear that canonically $T_p\partial M \subset T_pM$ and

$$T_p^+M \cap T_p^-M = T_p\partial M.$$

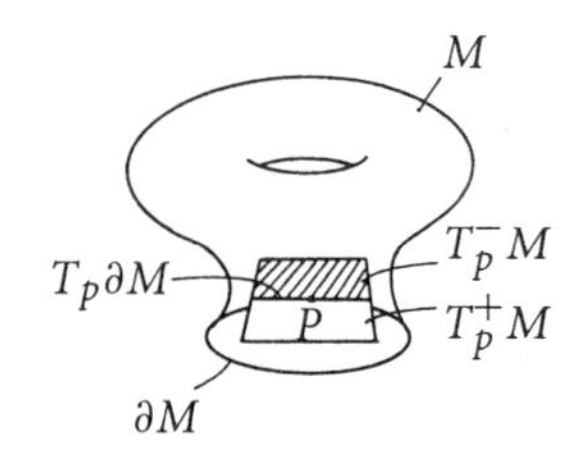

Figure 6.13. The half-spaces $T_p^{\pm}M$ for $p \in \partial M$

The elements of $T_p^- M \setminus T_p\partial M$ are called ***inward-pointing*** tangent vectors and those of $T_p^+ M \setminus T_p\partial M$ ***outward-pointing*** tangent vectors. A vector $v \in T_pM$ points inward (resp. outward) if and only if its first component v^1 is negative (resp. positive) relative to some (hence every) chart.

6.8 The Orientation Convention

Orientation and ***orienting atlas*** are defined for manifolds-with-boundary exactly as for ordinary manifolds. It is easy to see that the boundary of an *oriented* manifold M is *orientable* in any case, but this does not mean that ∂M is already canonically *oriented*. For that, we need a convention.

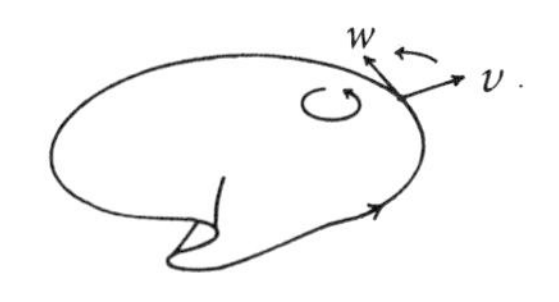

Figure 6.14. The orientation convention

Orientation convention. If M is an oriented n-dimensional manifold-with-boundary and $p \in M$, then a basis $w_1, \ldots, w_{n-1}$ of $T_p\partial M$ is said to be positively oriented (or, in the case $n = 1$, to have orientation $+1$) if and only if for some (hence every) outward-pointing vector v the basis $(v, w_1, \ldots, w_{n-1})$ of T_pM is positively oriented. From now on, let the boundary ∂M of an oriented manifold always be provided with this orientation.

Thus, if we use the right-hand rule to orient a three-dimensional submanifold-with-boundary, say a ball or a solid torus, of the real physical space that surrounds us, then when viewed from the outside the bounding surface has the counterclockwise orientation.

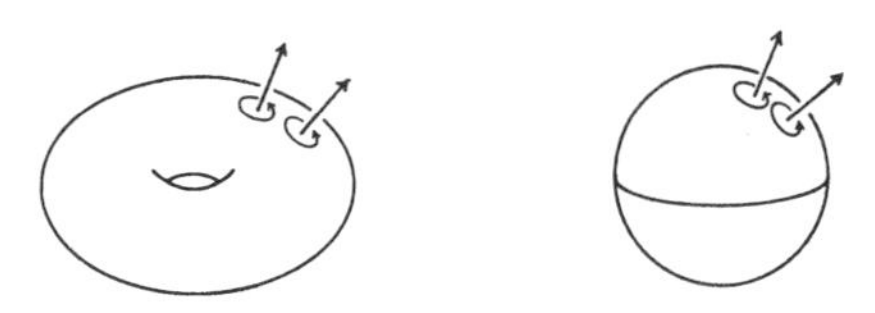

Figure 6.15. The orientation convention and the right-hand rule for objects in physical space

Since we have tangent spaces available for manifolds-with-boundary (and, on chart domains, coordinate vector fields $\partial_1, \dots, \partial_n$ as well), it is clear what is meant by k-forms ω on a manifold-with-boundary, when such forms are continuous or differentiable, what the vector space $\Omega^k M$ of differentiable k-forms on M is, and, finally, when an n-form ω on an n-dimensional manifold-with-boundary is integrable and what the integral $\int_M \omega$ means. All this brings us another step closer to Stokes's theorem.

6.9 Test

(1) Which of the the following are open in the topology of the half-space $\mathbb{R}^n_- = \{x \in \mathbb{R}^n : x^1 \leq 0\}$?

☐ $X := \{x \in \mathbb{R}^n : \|x\| < 1 \text{ and } x^1 < 0\}$.

☐ $X := \{x \in \mathbb{R}^n : \|x\| < 1 \text{ and } x^1 \leq 0\}$.

☐ $X := \{x \in \mathbb{R}^n : \|x\| \leq 1 \text{ and } x^1 \leq 0\}$.

(2) Let U be the part of the open square $(-1, 1) \times (-1, 1)$ that lies in the left half-plane $\mathbb{R}^2_-$. Thus

$$U = \{(x, y) \in \mathbb{R}^2 : -1 < x \leq 0 \text{ and } -1 < y < 1\}.$$

Let A denote the right side of U, and B the union of the other three sides. More precisely,

$$A := \{0\} \times (-1, 1),$$
$$B := \{-1\} \times [-1, 1] \cup [-1, 0] \times \{\pm 1\}.$$

As a subset of the topological space $\mathbb{R}^2_-$, U also has a *topological boundary*, namely the set $\dot{U}_{\mathbb{R}^2_-}$ of points of $\mathbb{R}^2_-$ that are neither interior nor exterior points of U. Similarly, we can consider $\dot{U}_{\mathbb{R}^2}$. This question focuses on the differences, such as they are, among ∂U, $\dot{U}_{\mathbb{R}^2_-}$, and $\dot{U}_{\mathbb{R}^2}$. Which of the following is true?

☐ $\partial U = A \cup B$, $\quad \dot{U}_{\mathbb{R}^2_-} = A \cup B$, $\quad \dot{U}_{\mathbb{R}^2} = A \cup B$.

□ $\partial U = A, \quad \dot{U}_{\mathbb{R}^2_-} = A \cup B, \quad \dot{U}_{\mathbb{R}^2} = A \cup B.$

□ $\partial U = A, \quad \dot{U}_{\mathbb{R}^2_-} = B, \quad \dot{U}_{\mathbb{R}^2} = A \cup B.$

(3) Let M denote a manifold-with-boundary. Can $M \setminus \partial M$ be compact if $\partial M \neq \emptyset$?

□ No, because then $M \setminus \partial M$ would also be closed, so ∂M would be open in M.

□ Yes. This happens if and only if M is compact.

□ Yes. By the Heine-Borel theorem this holds, for example, for all closed bounded submanifolds-with-boundary of $\mathbb{R}^n$.

(4) Can a zero-dimensional submanifold M_0 of a manifold-with-boundary M "touch" the boundary ($\overline{M}_0 \cap \partial M \neq \emptyset$)?

□ No, because M_0 consists of isolated points in $M \setminus \partial M$.

□ Yes. Let $M := \mathbb{R}^1_+$ and $M_0 := \{1/n : n = 1, 2, \dots\}$.

□ No, because zero-dimensional submanifolds are automatically closed. Hence $\overline{M}_0 \cap \partial M = M_0 \cap \partial M = \emptyset$ by definition.

(5) Let M be a manifold-with-boundary and $p \in M$. Is $M \setminus \{p\}$ a submanifold-with-boundary of M, with $\partial(M \setminus \{p\}) = \partial M \setminus \{p\}$?

□ Yes. *Every* open subset X of M is a submanifold-with-boundary, with $\partial X = X \cap \partial M$.

□ No. $M \setminus \{p\}$ is a submanifold-with-boundary, but $\partial(M \setminus \{p\}) = \partial M$ whenever $\dim M > 0$ because then $M \setminus \{p\}$ is dense in M.

□ Yes. The charts (U, h) on M with $p \notin U$ form an atlas on $M \setminus \{p\}$.

(6) Which of the following implications about connectedness are valid for manifolds-with-boundary M ?

- ☐ M is connected $\Longleftrightarrow M \setminus \partial M$ is connected.
- ☐ M is connected $\Longrightarrow \partial M$ is connected.
- ☐ ∂M is connected $\Longrightarrow M$ is connected.

(7) This question deals with cutting a manifold along a codimension-1 submanifold. Let M be an n-dimensional manifold without boundary, and let $M_0 := f^{-1}(c) \neq \emptyset$ be the preimage of a regular value c of a differentiable function f. Then M is the union of the n-dimensional submanifolds-with-boundary $A := f^{-1}([c, \infty))$ and $B := f^{-1}((-\infty, c])$, whose intersection is their common boundary M_0. You may visualize, for instance, that cutting along M_0 would split M into the disjoint union of A and B.

Now don't assume that a function f is given, but just a codimension-one closed nonempty submanifold (without boundary) $M_0 \subset M$. What happens if we "cut" M along M_0? To pose the question more precisely: Is M the union of two submanifolds-with-boundary A and B, with $\partial A = \partial B = A \cap B = M_0$? We would then say that M *splits when cut along* M_0. When does this happen?

- ☐ Not always. Let a circle be cut "along" a point, say, or a torus along a meridian.
- ☐ But there is always an open neighborhood X of M_0 in M that splits when cut along M_0. X need only be chosen to fit closely enough around M_0.
- ☐ This isn't true, either. Let a Möbius strip be cut along its "core" (midline), for instance, or $\mathbb{RP}^2$ along $\mathbb{RP}^1$: then no X splits.

(8) Let M be a manifold without boundary and let $X \subset M$ be open. Is the closure $\overline{X} \subset M$ a submanifold-with-boundary?

- ☐ No. $M = \mathbb{R}^3$, $X = \{(x, y, z) : x^2 + y^2 + z^2 > 0\}$ is a counterexample.
- ☐ No. $M = \mathbb{R}^3$, $X = \{(x, y, z) : x^2 + y^2 - z^2 > 1\}$ is a counterexample.

☐ No. $M = \mathbb{R}^3$, $X = \{(x, y, z) : x^2 + y^2 > z^2\}$ is a counterexample.

(9) Is ∂M always a set of measure zero in M?

☐ Yes, because $\{0\} \times \mathbb{R}^{n-1}$ is a set of measure zero for Lebesgue measure on $\mathbb{R}^n_-$.

☐ No. For example, the surface of the ball has measure $4\pi r^2 \neq 0$.

☐ No, only when $\partial M = \emptyset$.

(10) Let M be an oriented manifold without boundary. $M_1 := \{1\} \times M$ and $M_0 := \{0\} \times M$ are called the top and bottom, respectively, of the cylinder $[0, 1] \times M$ over M. Let them both be oriented as copies of M, in other words so that the canonical maps $M_1 \cong M \cong M_0$ are orientation-preserving. Now let the interval $[0, 1]$ be oriented as usual. Then our orientation convention induces the following orientation on the boundary:

☐ $\partial([0, 1] \times M) = M_0 + M_1$.

☐ $\partial([0, 1] \times M) = M_0 - M_1$.

☐ $\partial([0, 1] \times M) = M_1 - M_0$.

6.10 Exercises

EXERCISE 6.1. Let M be a manifold-with-boundary. Show that ∂M is closed in M.

EXERCISE 6.2. Let $f : M \to \mathbb{R}$ be a differentiable function that is regular everywhere on the compact manifold-with-boundary M. Show that f assumes its extrema on the boundary.

EXERCISE 6.3. Compact manifolds without boundary are called *closed*, and two closed manifolds M_0 and M_1 are called *bordant* if $M_0 + M_1$ is (diffeomorphic to) the boundary of a compact manifold-with-boundary. Prove that if M is closed and a, b are regular values of $f : M \to \mathbb{R}$, then $f^{-1}(a)$ and $f^{-1}(b)$ are bordant.

EXERCISE 6.4. Prove that every closed manifold M on which there exists a fixed-point-free differentiable involution τ is "null-bordant"; that is, it bounds a compact manifold.

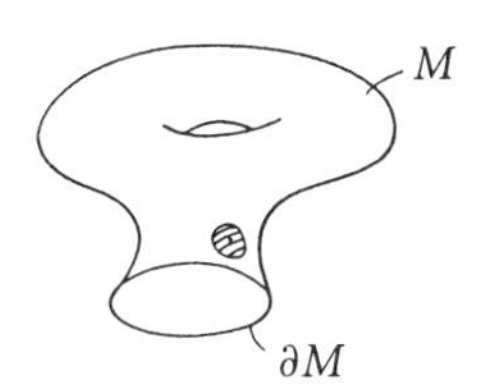

Figure 6.16.

6.11 Hints for the Exercises

FOR EXERCISE 6.1. This is intuitively clear: every point in $M \setminus \partial M$ has a neighborhood that does not meet the boundary. For the proof, you just need to make proper use of the relative topology of the half-space $\mathbb{R}^n_-$.

FOR EXERCISE 6.2. Incidentally, a regular function might not have an extremum on $M \setminus \partial M$ for a noncompact manifold, either, because it might not have any extrema at all. But, as you know, a continuous function on a compact topological space always has a maximum and a minimum. The exercise is so simple that no suitable hint occurs to me. Perhaps I should remind you that $f : M \to \mathbb{R}$ is regular if and only if $df_p \neq 0$ for all p.

FOR EXERCISE 6.3. The exercise is intended to make you *use* the lemma about $f^{-1}((-\infty, c])$ that was stated (with nothing but an allusion to the regular value theorem) at the end of Section 6.6.

FOR EXERCISE 6.4. This exercise is a bit harder than the other three. The not so obvious idea is (more or less) to connect each x and $\tau(x)$ by a line segment, so as to construct a compact manifold W with $\partial W = M$. But how can this be carried out technically? One can, for instance, start with the manifold (without boundary) $M \times \mathbb{R}$ and then take a quotient (also without boundary) $(M \times \mathbb{R})/\sim$ under a suitable free involution, as described in Section 1.6. This can be done in such a way that the desired W appears as a submanifold-with-boundary $f^{-1}((-\infty, c])$ in the quotient.

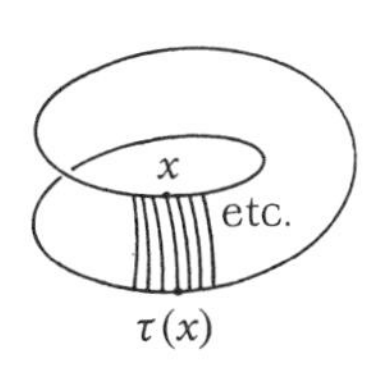

Figure 6.17.

The Intuitive Meaning of Stokes's Theorem

7.1 Comparison of the Responses to Cells and Spans

The actual definition of the Cartan (or exterior) derivative $d : \Omega^k M \to \Omega^{k+1} M$ will be postponed until the next chapter, and the proof of Stokes's theorem that $\int_M d\omega = \int_{\partial M} \omega$ until the chapter after that. In the present chapter I'll try to sketch how one could intuitively come up with the idea of the exterior derivative and conjecture Stokes's theorem.

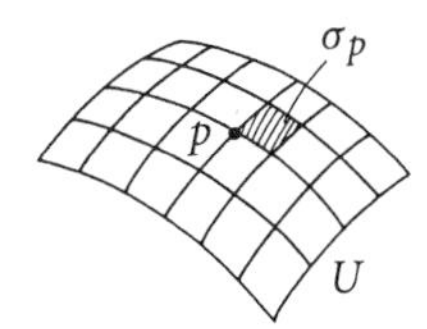

Figure 7.1.

We imagined a piece U of an oriented manifold as decomposed into small cells and the integral $\int_U \omega$ as the sum of the responses of the n-form ω to the cells, with each cell σ_p approximated by the tangential span $s_p = \text{span}(\Delta x^1 \partial_1, \ldots, \Delta x^n \partial_n)$. We can now look back at how the integral was formally introduced in Chapter 5 and estimate how well $\sum_p \omega_p(s_p)$ approximates $\int_U \omega$. If $a = \omega_{1\ldots n} \circ h^{-1}$ is the downstairs component function, then the actual contri-

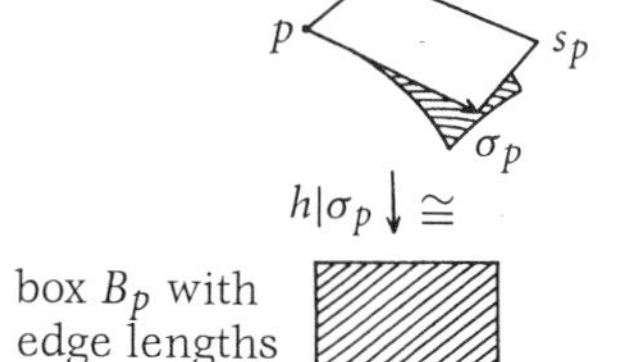

Figure 7.2. The box corresponds to the cell under h and to the span under dh_p.

bution of the cell to the integral is

$$\int_{\sigma_p} \omega = \int_{B_p} a(x)dx,$$

while its approximation

$$\omega_p(s_p) = \int_{B_p} a(h(p))dx$$

is the integral of the constant value $a(h(p)) = \omega_p(\partial_1, \ldots, \partial_n)$ and therefore equals $\omega_p(\partial_1, \ldots, \partial_n)\Delta x^1 \cdot \ldots \cdot \Delta x^n$. So if ω is a *continuous* n-form, for instance, then the absolute value of the error can be at most $\varepsilon_p \cdot \mathrm{Vol}(B_p)$, where ε_p denotes the variation of a on the box B_p; more precisely,

$$\varepsilon_p := \sup_{x \in B_p} |a(x) - a(h(p))|.$$

Hence the absolute value of the *total error* over the whole region U is less than or equal to $\max_p \varepsilon_p \cdot \mathrm{Vol}(h(U))$, and $\max_p \varepsilon_p$ becomes arbitrarily small for continuous ω when the grid is fine enough.

For continuous ω, this reasoning also shows how we can recover the alternating n-form $\omega_p \in \mathrm{Alt}^n T_p M$ from the integrals $\int_{\sigma_p} \omega$. For fixed orientation-preserving charts h, if we consider the edge lengths $\Delta x^1, \ldots, \Delta x^n$ of the cell at p as variables, then

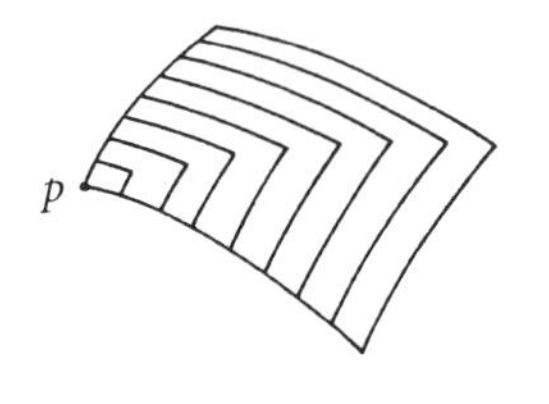

Figure 7.3. Cells σ_p with edge lengths Δx^μ getting smaller and smaller

$$\omega_p(\partial_1, \ldots, \partial_n) = \lim_{\Delta x \to 0} \frac{1}{\Delta x^1 \cdot \ldots \cdot \Delta x^n} \int_{\sigma_p} \omega.$$

This formula makes precise the statement that ω_p is the infinitesimal version at p of the integral of ω.

7.2 The Net Flux of an n-Form through an n-Cell

Stokes's theorem makes a statement about $(n-1)$-forms $\omega \in \Omega^{n-1}M$ on an oriented n-dimensional manifold. By its very nature, such a form responds to oriented tangential $(n-1)$-spans. But it also responds to oriented $(n-1)$-*cells*: approximately, through an approximating span; precisely, through integration over the cell as an $(n-1)$-dimensional manifold.

"Flux densities," which are described by 2-forms in three-dimensional space, give an intuitive picture of $(n-1)$-forms on oriented n-dimensional manifolds. The response of a flux density ω to an oriented 2-cell shows how much "flows through" the cell per unit time. The orientation lets us give opposite signs to the two possible directions of passage through the cell. It is a useful exercise to work out intuitively why such a flux density for infinitesimal cells is multilinear and alternating.

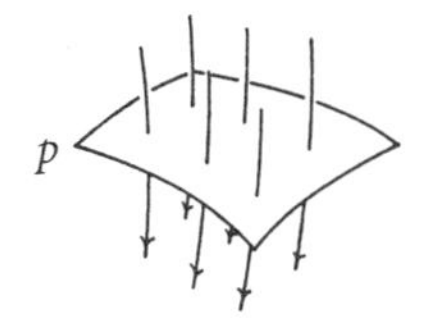

Figure 7.4.

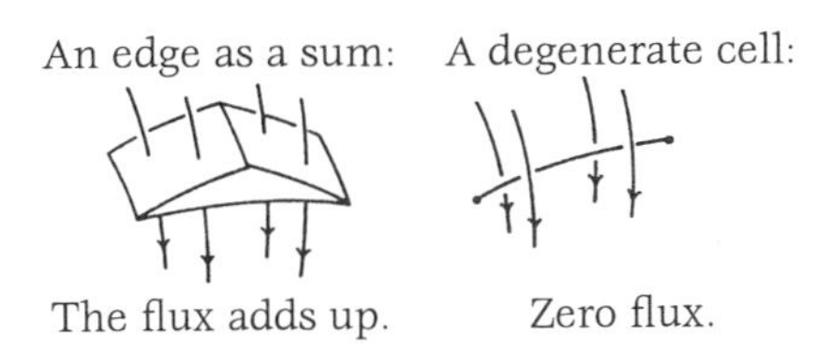

Figure 7.5. Flux densities are multilinear and alternating.

This intuitive picture suggests an interesting possibility for letting an $(n-1)$-form act on n-cells and then infinitesimally on n-spans. The "boundary" $\partial\sigma_p$ of an n-cell σ_p consists of $2n$ boundary cells of dimension $(n-1)$, a "front" and a "back" for each coordinate. We orient these $2n$ sides according to the same convention we used for the boundary of a manifold: The outward-pointing normal followed by the orientation of the boundary cell gives the orientation of σ. Now we can add the $2n$ responses *that ω gives to the oriented boundary cells* and thus define how ω should act on oriented n-cells, namely by

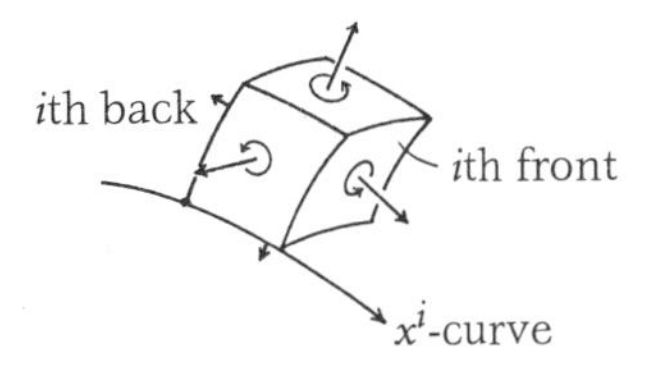

Figure 7.6. The $2n$ oriented boundary cells of an n-cell in an oriented n-dimensional manifold

$$\sigma \longmapsto \int_{\partial\sigma} \omega.$$

To make the notation unambiguous, we now write down explicitly the convention we have already used implicitly.

Notation. Let ω be a k-form on an n-dimensional manifold M and let $M_0 \subset M$ be an oriented k-dimensional submanifold or, in the case $k = n - 1$, the boundary ∂M of M, provided with an orientation. If $\iota : M_0 \hookrightarrow M$ denotes inclusion and $\iota^*\omega$ the induced k-form on M_0, we write

$$\int_{M_0} \iota^*\omega =: \int_{M_0} \omega.$$

The suppression of ι^* is justified, first because

$$(\iota^*\omega)|_p(v_1, \ldots, v_k) = \omega_p(v_1, \ldots, v_k),$$

and second because there is no risk of confusion with the notation $\int_{M_0} \omega := \int_M \omega_{M_0}$ of Section 5.3: for $k < n$, a k-form can't be integrated over M anyway, so $\int_M \omega_{M_0}$ would make no sense, and for $k = n$ it's really the same thing.

So much for notation. The intuitive meaning of $\int_{\partial\sigma} \omega$ is the *net flux* through the n-cell σ! What we measure with $\int_{\partial\sigma} \omega$ is what flows out of the n-cell σ per unit time, since the orientation of $\partial\sigma$ given here assigns incoming flow a negative value and outgoing flow a positive value. Thus the excess $\int_{\partial\sigma} \omega$ can be called the *source strength* of σ.

Stokes's theorem is ultimately based on this idea of balancing the incoming and outgoing flows. When we resume our discussion from Section 5.1 about the relative merits of densities and forms, we have to note that a treatment of the net flux using *densities* would in any case require a notion of "$(n-1)$-densities" that would take into account the orientation of the cells of $\partial\sigma$, for without some distinction between incoming and outgoing flow there can be no net flow. The k-forms are already set up for this.

7.3 Source Strength and the Cartan Derivative

Now, is this source strength of an $(n-1)$-form ω, viewed as a correspondence $\sigma \to \int_{\partial\sigma} \omega$, really the action of a *differential form* of degree n? In other words, for every $(n-1)$-form ω is there an n-form η such that $\int_\sigma \eta = \int_{\partial\sigma} \omega$ for oriented n-cells? If so, then, as we saw earlier, this "source density" η would at least have to satisfy

$$\eta(\partial_1, \dots, \partial_n) = \lim_{\Delta x \to 0} \frac{1}{\Delta x^1 \cdot \ldots \cdot \Delta x^n} \int_{\partial\sigma} \omega,$$

and this already reveals one way of trying to answer the question: Check whether this limit exists and, if it does, whether η is independent of the choice of chart. Then prove that $\int_\sigma \eta = \int_{\partial\sigma} \omega$ for the η so defined. Actually, we have no intention of taking this approach because we'll reach our goal in a more elegant, though more formal, way. But if we imagine ourselves in a fictitious pioneer period of the Cartan calculus, this path is definitely the right one, and it leads to the insight that for every $\omega \in \Omega^{n-1}M$ there is in fact exactly one n-form ***that responds to oriented n-cells as ω itself responds to their boundaries***. This n-form is called the ***Cartan derivative*** of ω and is denoted by $d\omega$.

Incidentally, if we consider the contribution of the ith pair of sides to the limit

$$d\omega(\partial_1, \dots, \partial_n) = \lim_{\Delta x \to 0} \frac{1}{\Delta x^1 \cdot \ldots \cdot \Delta x^n} \int_{\partial\sigma} \omega,$$

we not only *obtain* the formula

$$d\omega(\partial_1, \dots, \partial_n) = \sum_{i=1}^{n} (-1)^{i-1} \frac{\partial}{\partial x^i} \omega(\partial_1, \dots, \widehat{\imath}, \dots, \partial_n)$$

for the Cartan derivative in coordinates, we also understand the intuitive meaning of its individual summands.

7.4 Stokes's Theorem

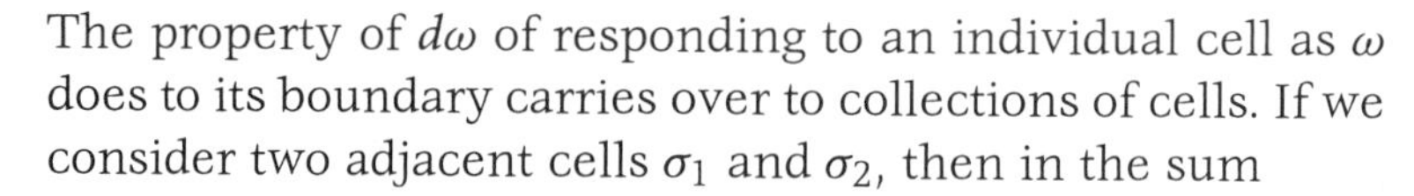

The property of $d\omega$ of responding to an individual cell as ω does to its boundary carries over to collections of cells. If we consider two adjacent cells σ_1 and σ_2, then in the sum

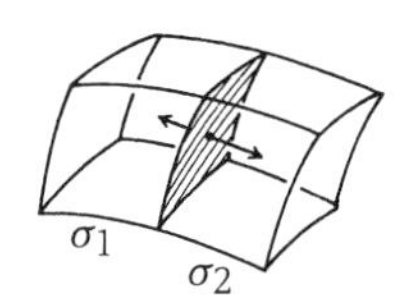

Figure 7.7.

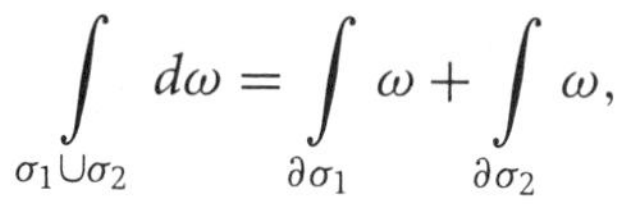

$$\int_{\sigma_1 \cup \sigma_2} d\omega = \int_{\partial\sigma_1} \omega + \int_{\partial\sigma_2} \omega,$$

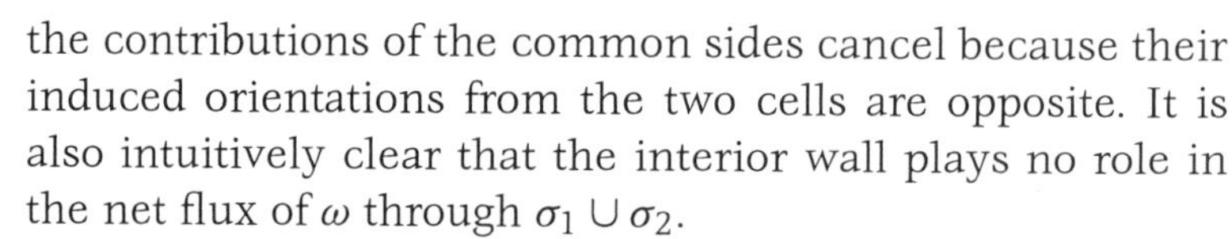

the contributions of the common sides cancel because their induced orientations from the two cells are opposite. It is also intuitively clear that the interior wall plays no role in the net flux of ω through $\sigma_1 \cup \sigma_2$.

If we now think of a compact oriented manifold-with-boundary as a single collection of cells, we see that the contributions of the interior sides of the cells all cancel in the sum

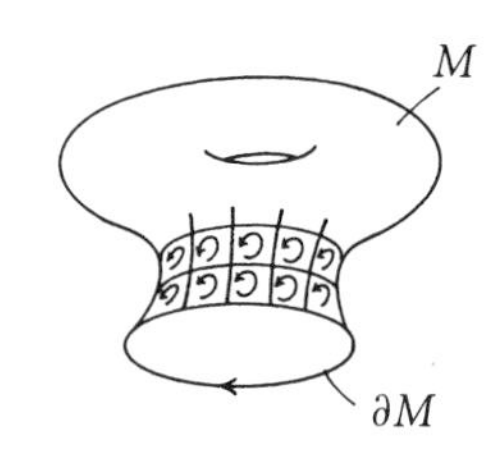

Figure 7.8. $\int_\sigma d\omega = \int_{\partial\sigma} \omega$ for cells (definition of d), so $\int_M d\omega = \int_{\partial M} \omega$ (Stokes's theorem).

$$\int_M d\omega = \sum_p \int_{\sigma_p} d\omega = \sum_p \int_{\partial\sigma_p} \omega$$

and only the integrals over the sides that form the boundary ∂M are left, so that

$$\int_M d\omega = \int_{\partial M} \omega.$$

This is Stokes's theorem.

As we said earlier, this is not how we'll actually prove Stokes's theorem. Decomposing the whole manifold into a grid of cells in a rigorous way would be a technically demanding project, to say nothing of its being impossible in general unless one also allowed certain "singular cells," such as occur, for instance, in angular coordinates on S^2 at the poles.

Although the idea of decomposition into cells does not lead to an elegant proof, it describes the geometric content of the theorem extremely well—in fact, it reduces the theorem at the intuitive level to a truism.

7.5 The de Rham Complex

The definition of the Cartan derivative $d\omega$ through the action of ω on the boundary is not limited to $(n-1)$-forms. For any differential k-form $\omega \in \Omega^k M$, with k arbitrary, there is exactly one $(k+1)$-form $d\omega \in \Omega^{k+1} M$ that responds to oriented $(k+1)$-cells as ω does to their oriented boundaries. This yields a whole *sequence*

$$0 \to \Omega^0 M \xrightarrow{d} \Omega^1 M \xrightarrow{d} \cdots \xrightarrow{d} \Omega^{n-1} M \xrightarrow{d} \Omega^n M \to 0$$

of linear maps.

The Cartan derivative $d : \Omega^0 M \to \Omega^1 M$ on *zero-forms* (C^∞ functions on M) is simply the differential: For an oriented 1-cell σ as in Figure 7.9, q and p are positively and negatively oriented, respectively, by the orientation convention. Thus $\int_\sigma d\omega = \omega(q) - \omega(p)$ for $\omega \in \Omega^0 M$, and there is no clash between our previous notation $df \in \Omega^1$ for the differential of a function and our notation d for the Cartan derivative.

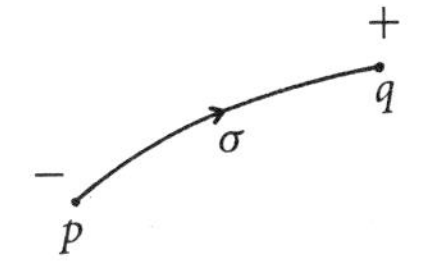

Figure 7.9.

The sequence of Cartan derivatives is what in homological algebra is called a *complex*; that is, $d \circ d = 0$. More precisely, if $\omega \in \Omega^{k-1} M$ and σ is an oriented $(k+1)$-cell, then

$$\int_\sigma dd\omega = \int_{\partial\sigma} d\omega = \int_{\partial\partial\sigma} \omega,$$

where the integral over $\partial\partial\sigma$ just denotes the sum of the integrals over the sides of the sides of σ. But in this sum the integral is taken twice, with opposite orientations, over each edge. Hence $\int_{\partial\partial\sigma} \omega = 0$. Or: If we dare to think of the $(k+1)$-cell σ, despite its edges and corners, as a manifold-with-boundary (as we may, if all we want to do is integrate over it), then as a manifold without boundary $\partial\sigma$ has empty boundary $\partial\partial\sigma = \emptyset$, and applying Stokes's theorem twice gives $\int_\sigma dd\omega = \int_{\partial\sigma} d\omega = \int_\emptyset \omega = 0$ since an integral over the empty manifold is of course zero. In any case, we understand the property $dd = 0$ as a consequence of the geometric fact

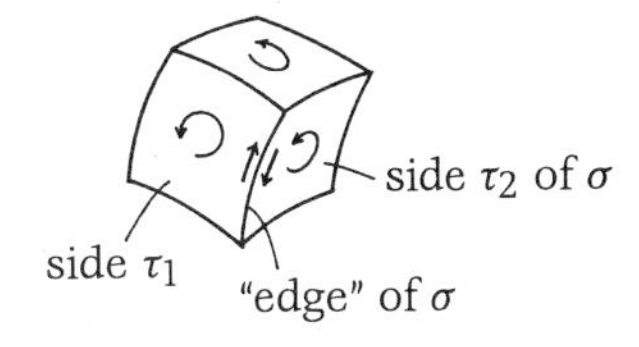

Figure 7.10. The response $\int_\sigma dd\omega$ of $dd\omega$ to a $(k+1)$-cell is zero.

"$\partial\partial = 0$." The complex

$$0 \to \Omega^0 M \xrightarrow{d} \Omega^1 M \xrightarrow{d} \cdots \xrightarrow{d} \Omega^{n-1} M \xrightarrow{d} \Omega^n M \to 0$$

is called the ***de Rham complex*** of M.

7.6 Simplicial Complexes

The de Rham complex canonically defines a contravariant functor from the differentiable category to the category of (cochain) complexes and represents an important interface between analysis and algebraic topology.

Of course, to explain this in the technically correct sense would first require an introduction to algebraic topology and would therefore go beyond the scope of the present book. But I'll try to give you an intuitive idea. To do this, I have to start by telling you about a completely different kind of complex.

"Complex" is a general word for something assembled from individual building blocks. It has lost this naive meaning in the expression *de Rham complex*, but retains it in the phrase *simplicial complex*. Imagine that you were allowed to assemble arbitrary things from (closed) tetrahedra, triangles, line segments, and points in $\mathbb{R}^3$ as three-, two-, one-, and zero-dimensional building blocks ("simplices"), where you had to follow only two rules of play:

(1) You may use only finitely many building blocks at a time.

(2) Adjacent building blocks must be mutually compatible. More precisely: the intersection of any two building blocks must be either empty or a common subsimplex.

The subsimplices of a tetrahedron, for example, are its vertices, edges, and faces. The rules are similar in $\mathbb{R}^n$, where analogous building blocks up to dimension n are possible and permitted. The things you can assemble by following these rules are called *finite simplicial complexes*. With some precautions ("locally finite" instead of "finite" in the first rule of play), infinitely many building blocks can also be allowed.

If you're not deliberately looking for counterexamples, you can imagine a simplicial building-block model for any geometric object you encounter in $\mathbb{R}^n$: the ball, the cone, and the torus; any kind of manifold or non-manifold—usually not quite genuine, because they'll have corners and edges, but still *homeomorphic* to the object and hence faithfully describing its *topological* properties.

To get our hands on such topological properties of the original, we now consider *simplicial chains* in the model. Anyone can picture a "chain" of finitely many oriented edges of building blocks that runs from one vertex of the simplicial complex to another, whose "boundary" thus consists of the (positively oriented) terminal point and the (negatively oriented) initial point. If the initial point coincides with the terminal point, the chain is a "cycle." Obvious terminology!

But if we want to make the union of chains into the operation of an abelian group—and not just stay in one dimension—we are automatically led to the following generalization of the concept of chains.

Definition. The k-dimensional ***simplicial chains*** of a simplicial complex X are described by formal linear combinations

$$\lambda_1\sigma_1 + \cdots + \lambda_r\sigma_r$$

(where the coefficients λ_i are integers) of oriented k-dimensional (sub-) building blocks of the simplicial complex and added accordingly, with the provision that a k-simplex σ goes to $-\sigma$ under a change of orientation.

The k-dimensional chains of X thus form an abelian group $S_k(X)$; each individual oriented k-simplex σ has a $(k-1)$-chain $d\sigma$ as boundary (with the same orientation convention as for manifolds-with-boundary), and this also defines a boundary chain $dc \in S_{k-1}(X)$ for each *k-chain* $c \in S_k(X)$. A chain c with $dc = 0$ is called a ***cycle***, and the sequence

$$0 \to S_n(X) \xrightarrow{d} S_{n-1}(X) \xrightarrow{d} \cdots \xrightarrow{d} S_1(X) \xrightarrow{d} S_0(X) \to 0$$

of boundary operators is called the ***simplicial chain complex*** of X.

The boundary of the boundary chain of a k-simplex is clearly zero since, just as for a cell, the contribution of each side cancels with the oppositely oriented contributions of the adjacent sides. Hence $d \circ d = 0$ for chains as well. In words: *All boundaries are cycles.*

But not all cycles are boundaries. A meridian cycle on a simplicial torus, for instance, doesn't look as though it could be the boundary of a 2-chain. And it is precisely these *nonbounding* cycles that seem to say something about the topological structure of the simplicial complex and hence also about the structure of the geometric object that really interests us, for which the simplicial complex is only the building-block model. But how can we get hold of this information mathematically?

If we want to suppress the uninteresting boundaries in calculations, we have to compute with cycles "modulo boundaries"; that is, we define cycles as equivalent, or *homologous*, if they differ only by a boundary. The equivalence classes, or *homology classes*, of k-cycles are then the elements of the kth *homology group* of X, the quotient of the group of cycles by the group of boundaries.

Definition. If X is a simplicial complex, the abelian group

$$H_k(X, \mathbb{Z}) := \frac{\ker(d : S_k(X) \to S_{k-1}(X))}{\operatorname{im}(d : S_{k+1}(X) \to S_k(X))}$$

is called the ***kth simplicial homology group*** of X.

If X is a *finite* simplicial complex, for instance, then by construction $H_k(X, \mathbb{Z})$ is a finitely generated abelian group, which in principle can be computed by elementary means.

But does it really tell us anything about the original geometric object, or is it influenced by the uninteresting details of how we constructed the building-block model?

If it told us nothing, we would probably not be discussing this hundred-or-so-year-old invention today. A method called *simplicial approximation* can be used to show not only that homeomorphic simplicial complexes have isomorphic homology groups, but even that simplicial homology canonically defines a functor from the category of "triangulable" topological spaces (those that are homeomorphic to a simplicial complex) and continuous maps to the category of abelian groups (which are *graded* by the index k).

With this, homology theory was established.

7.7 The de Rham Theorem

The success of homology theory was striking. Famous old theorems shrank to little lemmas, and masses of unsuspected new results were proved. Applying the homology functor gave, so to speak, an X-ray look inside apparently impenetrable geometric problems.

You can imagine that these advances were accompanied by improvements in the methods. What crystallized out of this as the essential rule for success was to assign *chain complexes*

$$\cdots \xrightarrow{d} C_{k+1}(X) \xrightarrow{d} C_k(X) \xrightarrow{d} C_{k-1}(X) \xrightarrow{d} \cdots$$

to geometric objects X in a natural, functorial way. These chain complexes are sequences of homomorphisms between algebraic objects—abelian groups for instance, or vector spaces or modules over rings—that satisfy the condition $d \circ d = 0$ and whose kth *homology*

$$H_k(C(X), d) := \frac{\ker(d : C_k(X) \to C_{k-1}(X))}{\operatorname{im}(d : C_{k+1}(X) \to C_k(X))}$$

can therefore be studied. It became clear, for example, that simplicial homology is independent of the triangulation. So shouldn't it be possible to define it directly, without resorting to a building-block model, and then define it in complete generality for arbitrary topological spaces? *Singular homology* was discovered as a solution to this problem and has turned out to be pivotal.

A *singular k-simplex* of a topological space is just a continuous map $\sigma : \Delta_k \to X$ from the k-dimensional standard simplex to X—when $k = 1$ this is a continuous path in X—and the k-chains of this theory are formal linear combinations with integer coefficients of singular k-simplices. The resulting *singular homology groups* $H_k(X, \mathbb{Z})$ can no longer be computed directly from the definitions, but the developing homology theory had already left naive computational methods behind in any case and replaced them by more elegant axiomatic methods.

Of special significance in discovering new homology theories was the application of algebraic functors to chain complexes associated with tried and true theories. More information lies dormant in a chain complex than is extracted by homology. So one can hope to find something new by subjecting the chain complex to an algebraic manipulation *before* taking the homology quotients $\ker d / \operatorname{im} d$, as long as this manipulation preserves the property $d \circ d = 0$. For example, one can take an abelian group G and tensor all the "chain groups" $C_k(X)$ with it. In the case of singular homology, this leads to singular homology *with coefficients in* G, whose groups are denoted by $H_k(X, G)$.

A polished *algebraic* theory of chain complexes eventually became such a compelling technical necessity for homology theory, which was growing into an industry, that an independent new subdiscipline, *homological algebra*, was generated in its wake.

Of course, one of the algebraic functors that could be tested on existing chain complexes and actually *were* used early on is the Hom-functor $\mathrm{Hom}(-, G)$. Since it is contravariant, it turns a chain complex into what is called a *cochain complex*, whose grading now *ascends*. For instance, it turns the singular chain complex into the cochain complex with coefficients in G,

$$\cdots \overset{\delta}{\longleftarrow} \mathrm{Hom}(C_{k+1}(X), G) \overset{\delta}{\longleftarrow} \mathrm{Hom}(C_k(X), G) \overset{\delta}{\longleftarrow} \cdots .$$

The homology groups of this complex are logically called singular *co*homology groups with coefficients in G and written $H^k(X, G)$.

It was not immediately evident what a significant extension of homology theory had been stumbled upon. In fact, it was realized only gradually that for singular cohomology—in contrast to homology!—with coefficients in a commutative ring R, there is a product

$$\smile : H^r(X, R) \times H^s(X, R) \longrightarrow H^{r+s}(X, R),$$

the *cup product*, which turns the cohomology groups into a *cohomology ring*. This has far-reaching consequences.

As you can see, the de Rham complex is also a cochain complex and defines a cohomology theory for the category of manifolds. The cohomology groups $H^k_{\mathrm{dR}}M$ of this de Rham cohomology are real vector spaces, and with the wedge product they form a cohomology ring. This is outwardly quite similar to singular cohomology with coefficients in $\mathbb{R}$! But the origin of de Rham cohomology makes it seem exotic among the other homology theories, which can't deny their descent from simplicial homology. Its boundary homomorphism, the Cartan derivative, is a differential operator!

Georges de Rham was the first to discover the nature of this exotic cohomology theory. It is the real singular cohomology of manifolds, and the wedge product is the cup product.

Stokes's theorem makes the connection. More precisely, one can integrate a k-form ω on M over a (differentiable) singular k-simplex σ in M by setting

$$\int_\sigma \omega := \int_{\Delta_k} \sigma^*\omega.$$

Thus $\int_c \omega$ is also defined for (differentiable) singular k-chains, and applying Stokes's theorem to Δ_k (the vertices and edges cause no real difficulties) gives $\int_c d\eta = \int_{dc} \eta$. So we have linear maps

$$H^k_{\mathrm{dR}} M \longrightarrow \mathrm{Hom}(H_k^{\mathrm{diffb}}(M, \mathbb{Z}), \mathbb{R}) \longleftarrow H^k(M, \mathbb{R}),$$

the first just through integration over singular cycles and the second directly from the definition of singular cohomology. Both are isomorphisms: Methods of ordinary homology theory show this for the second map, but the statement that the integration map is an isomorphism is the crux of de Rham's theorem and not easy to prove.

De Rham's theorem turned out to be a momentous discovery. It brought to light the deep connections between analysis, the powerful, well-established discipline, and algebraic topology, the successful newcomer—connections that play a major role in present-day mathematics. I'm thinking of the Atiyah-Singer index theorem, for example, and its ramifications, which extend even into theoretical physics.

In a more elementary way, the de Rham complex is an everyday presence in classical vector analysis—wherever the three familiar differential operators gradient, curl, and divergence appear in the three-dimensional physical space M. As we have yet to see in detail (in Chapter 10), they correspond exactly to the three Cartan derivatives:

$$0 \to \Omega^0 M \xrightarrow[\mathrm{grad}]{d} \Omega^1 M \xrightarrow[\mathrm{curl}]{d} \Omega^2 M \xrightarrow[\mathrm{div}]{d} \Omega^3 M \to 0.$$

This is why, for instance, the divergence of a curl and the curl of a gradient are always zero, and why statements about the de Rham complex always also have, in passing, a direct interpretation in classical vector analysis.

The present chapter, exactly in the middle of the book, is intended to convey something different from what can be asked about in tests and exercises. Perhaps with a deeper understanding of our subject, we now return to its technical details.

8 CHAPTER: The Wedge Product and the Definition of the Cartan Derivative

8.1 The Wedge Product of Alternating Forms

To define the Cartan derivative we use a tool from multilinear algebra, the exterior, or "wedge," product of alternating multilinear forms.

Definition. Let V be a real vector space and let $\omega \in \mathrm{Alt}^r V$ and $\eta \in \mathrm{Alt}^s V$. Then the alternating $(r+s)$-form $\omega \wedge \eta \in \mathrm{Alt}^{r+s} V$ defined by

$$\omega \wedge \eta(v_1, \ldots, v_{r+s}) :=$$

$$\frac{1}{r!s!} \sum_{\tau \in \mathcal{S}_{r+s}} \operatorname{sgn} \tau \cdot \omega(v_{\tau(1)}, \ldots, v_{\tau(r)}) \cdot \eta(v_{\tau(r+1)}, \ldots, v_{\tau(r+s)})$$

is called the ***exterior***, or ***wedge, product*** of ω and η.

Each summand is already multilinear in the variables $v_1, \ldots, v_{r+s}$. The way the big alternating sum—as I'll call it because of $\operatorname{sgn} \tau$—is constructed guarantees that $\omega \wedge \eta$ is alternating. But many of the summands are repeated: each of the $r!s!$ permutations that produce the same partition

$$\{1, \ldots, r+s\} = \{\tau(1), \ldots, \tau(r)\} \cup \{\tau(r+1), \ldots, \tau(r+s)\}$$

into one subset of r elements and another of s elements also yields the same summands, precisely because ω and η are assumed to be alternating. Hence $(\omega \wedge \eta)(v_1, \ldots, v_{r+s})$ is also given by the well-defined sum

$$\sum_{[\tau] \in \mathcal{Z}_{r,s}} \operatorname{sgn} \tau \cdot \omega(v_{\tau(1)}, \ldots, v_{\tau(r)}) \cdot \eta(v_{\tau(r+1)}, \ldots, v_{\tau(r+s)})$$

over the $\binom{r+s}{r}$-element set $\mathcal{Z}_{r,s}$ of these partitions.

Lemma. *The wedge product $\wedge$ has the following properties:*

(1) *For every real vector space V, the wedge product turns the direct sum $\bigoplus_{k=0}^{\infty} \mathrm{Alt}^k V$ into a graded anticommutative algebra with identity. More precisely, the following hold for every r, s, $t \geq 0$:*

(i) *The wedge product $\wedge : \mathrm{Alt}^r V \times \mathrm{Alt}^s V \to \mathrm{Alt}^{r+s} V$ is bilinear.*

(ii) *The wedge product is associative, i.e., $(\omega \wedge \eta) \wedge \zeta = \omega \wedge (\eta \wedge \zeta)$ for $\omega \in \mathrm{Alt}^r V$, $\eta \in \mathrm{Alt}^s V$, and $\zeta \in \mathrm{Alt}^t V$.*

(iii) *The wedge product $\wedge$ is anticommutative; that is, $\eta \wedge \omega = (-1)^{r \cdot s} \omega \wedge \eta$ for $\omega \in \mathrm{Alt}^r V$ and $\eta \in \mathrm{Alt}^s V$.*

(iv) *The 0-form $1 \in \mathrm{Alt}^0 V = \mathbb{R}$ satisfies $1 \wedge \omega = \omega$ for all $\omega \in \mathrm{Alt}^r V$.*

(2) *The wedge product is "natural." In other words, it is compatible with linear maps: $f^*\omega \wedge f^*\eta = f^*(\omega \wedge \eta)$ for every linear map $f : W \to V$ and all $\omega \in \mathrm{Alt}^r V$, $\eta \in \mathrm{Alt}^s V$.*

Sketch of the Proof. Properties (i), (iv), and (2) follow trivially from the defining formula. Anticommutativity (iii) is also immediate. To verify associativity, think of $\omega \wedge \eta(v_1, \ldots, v_{r+s})$ as the sum over the partitions of $\{1, \ldots, r+s\}$ into one subset of r elements and another of s elements, as explained above. Then we see that $(\omega \wedge \eta) \wedge \zeta$ and $\omega \wedge (\eta \wedge \zeta)$, applied to $(v_1, \ldots, v_{r+s+t})$, are one and the same sum over the set $\mathcal{Z}_{r,s,t}$ of partitions of $\{1, \ldots, r+s+t\}$

into one subset of r elements, a second of s elements, and a third of t elements:

$$(\omega \wedge \eta \wedge \zeta)(v_1, \ldots, v_{r+s+t}) :=$$

$$\frac{1}{r!s!t!} \sum_{\tau \in \mathcal{S}_{r+s+t}} \operatorname{sgn} \tau \cdot \omega(v_{\tau(1)}, \ldots, v_{\tau(r)}) \cdot$$

$$\eta(v_{\tau(r+1)}, \ldots, v_{\tau(r+s)}) \cdot \zeta(v_{\tau(r+s+1)}, \ldots, v_{\tau(r+s+t)}). \qquad \square$$

Now we can state another, shorter, version of the lemma: *The wedge product turns $\bigoplus_{k=0}^{\infty} \mathrm{Alt}^k$ into a contravariant functor from the category of real vector spaces and linear maps to the category of real graded anticommutative algebras with identity and their homomorphisms.*

8.2 A Characterization of the Wedge Product

The wedge product is not yet characterized by these properties. For instance, if we chose real numbers $f(n) \neq 0$ for $n \geq 0$, requiring only that $f(0) = 1$, then conditions (1) and (2) would still hold for the wedge product $\tilde{\wedge}$ defined by

$$\omega \tilde{\wedge} \eta := \frac{f(r) f(s)}{f(r+s)} \omega \wedge \eta \quad \text{for } \omega \in \mathrm{Alt}^r V,\ \eta \in \mathrm{Alt}^s V.$$

But according to our definition, the wedge product also satisfies the following normalization condition.

Note. *Let $e_1, \ldots, e_k$ denote the canonical basis of $\mathbb{R}^k$ and $\delta^1, \ldots, \delta^k$ the corresponding dual basis of $\mathbb{R}^{k*} = \mathrm{Alt}^1 \mathbb{R}^k$. Then*
(3) $\delta^1 \wedge \ldots \wedge \delta^k(e_1, \ldots, e_k) = 1$ *for all $k \geq 1$.* $\square$

Theorem. *Only $\wedge$ satisfies* (1), (2), *and* (3).

PROOF. More precisely, the theorem says that if a binary operation $\wedge : \mathrm{Alt}^r V \times \mathrm{Alt}^s V \to \mathrm{Alt}^{r+s} V$ satisfies conditions (1)–(3) above for all V, r, s, then it coincides with the wedge product given explicitly in Section 8.1. Let $\wedge$ be an arbitrary such binary operation. Then the following also holds:

(4) *Let $e_1, \ldots, e_n$ be a basis of a real vector space V and $\delta^1, \ldots, \delta^n$ the dual basis, and let $1 \leq \nu_1 < \cdots < \nu_k \leq n$. Then*

$$\delta^{\mu_1} \wedge \ldots \wedge \delta^{\mu_k}(e_{\nu_1}, \ldots, e_{\nu_k}) = \begin{cases} \operatorname{sgn} \tau & \text{if } \tau \text{ exists,} \\ 0 & \text{otherwise,} \end{cases}$$

where τ denotes the permutation sending $\nu_1, \ldots, \nu_k$ to $\mu_1, \ldots, \mu_k$.

To prove (4), let V_0 denote the k-dimensional subspace of V spanned by $e_{\nu_1}, \ldots, e_{\nu_k}$ and $\iota : V_0 \hookrightarrow V$ the inclusion map. Then naturality (2) implies that

$$\delta^{\mu_1} \wedge \ldots \wedge \delta^{\mu_k}(e_{\nu_1}, \ldots, e_{\nu_k}) = \iota^*\delta^{\mu_1} \wedge \ldots \wedge \iota^*\delta^{\mu_k}(e_{\nu_1}, \ldots, e_{\nu_k}).$$

If $(\mu_1, \ldots, \mu_k)$ is *not* a permutation of $\nu_1 < \cdots < \nu_k$, then either $\mu_i = \mu_j$, for some $i \neq j$, in which case $\delta^{\mu_i} \wedge \delta^{\mu_j} = 0$ by anticommutativity (1) (iii), or there exists an i with $\mu_i \neq \nu_j$ for all j. But then we have $\iota^*\delta^{\mu_i} = 0$. On the other hand, if $(\mu_1, \ldots, \mu_k)$ does come from a permutation of $(\nu_1, \ldots, \nu_k)$, then $\mu_i = \nu_{\tau(i)}$ and

$$\iota^*\delta^{\mu_1} \wedge \ldots \wedge \iota^*\delta^{\mu_k} = \operatorname{sgn} \tau \cdot \iota^*\delta^{\nu_1} \wedge \ldots \wedge \iota^*\delta^{\nu_k}$$

by anticommutativity. Now (4) follows from the normalization condition (3) and naturality applied to $V_0 \cong \mathbb{R}^k$. What we have shown so far is that (1)–(3) $\Rightarrow$ (4).

With regard to our goal of proving the theorem, we have in particular the partial result that $\delta^{\mu_1} \wedge \ldots \wedge \delta^{\mu_k}$ is independent of the choice of the operation $\wedge$ satisfying (1)–(3). But to show this for arbitrary products $\omega \wedge \eta$, we have to write ω and η as linear combinations of such products of 1-forms. More precisely, we claim that (1)–(3) imply one more condition.

(5) *Let $\omega_{\mu_1 \ldots \mu_k} := \omega(e_{\mu_1}, \ldots, e_{\mu_k})$ be the components of the form $\omega \in \operatorname{Alt}^k V$ with respect to a basis $e_1, \ldots, e_n$ of V, and again let $\delta^1, \ldots, \delta^n$ denote the dual basis. Then*

$$\omega = \sum_{\mu_1 < \cdots < \mu_k} \omega_{\mu_1 \ldots \mu_k} \delta^{\mu_1} \wedge \ldots \wedge \delta^{\mu_k}.$$

To prove (5) we need only check that for $\nu_1 < \cdots < \nu_k$ both sides give the same result on $(e_{\nu_1}, \ldots, e_{\nu_k})$. But this follows directly from (4), and the proof of (5) is done.

Because of (5) and (4), we know now that for every finite-dimensional V the product $\omega \wedge \eta \in \mathrm{Alt}^{r+s}V$ is given by the operation defined explicitly in Section 8.1. But this suffices for the proof of the theorem, since naturality (2) implies that

$$(\omega \wedge \eta)(v_1, \ldots, v_{r+s}) = (\iota^*\omega \wedge \iota^*\eta)(v_1, \ldots, v_{r+s})$$

for arbitrary V, where $\iota : V_0 \hookrightarrow V$ denotes the inclusion of the finite-dimensional vector space V_0 into V. This proves the theorem. □

Note the following consequence of (5), which we state explicitly.

Corollary. *If* $(e_1, \ldots, e_n)$ *is a basis of* V *and* $(\delta^1, \ldots, \delta^n)$ *the dual basis, then* $(\delta^{\mu_1} \wedge \ldots \wedge \delta^{\mu_k})_{\mu_1 < \cdots < \mu_k}$ *is a basis of* $\mathrm{Alt}^k V$.

8.3 The Defining Theorem for the Cartan Derivative

This is enough for the time being about the wedge product as a concept from multilinear algebra. We now want to exploit it for analysis on manifolds. In what follows, manifolds may be manifolds-with-boundary if nothing is said otherwise.

Definition. Let M be a differentiable manifold. The ***wedge product***

$$\begin{aligned} \wedge : \Omega^r M \times \Omega^s M &\longrightarrow \Omega^{r+s} M, \\ (\omega, \eta) &\longmapsto \omega \wedge \eta \end{aligned}$$

of differential forms on M is defined pointwise in a natural way, by setting $(\omega \wedge \eta)_p := \omega_p \wedge \eta_p$ for every $p \in M$.

Observe that the wedge product with a 0-form (that is, with a function) is simply the ordinary product: $f \wedge \eta = f\eta$ for $f \in \Omega^0(M)$ by properties (1)(i),(iv), p. 134.

Note (see Section 8.1). *The wedge product turns* $\Omega^* := \bigoplus_{k=0}^{\infty} \Omega^k$ *into a contravariant functor from the category of manifolds and differentiable maps to the category of real graded anticommutative algebras with identity.*

Now let (U, h) be a chart. Recall (see the lemma in Section 3.5) that at every point of U the 1-forms $dx^1, \ldots, dx^n$ are the dual basis to the basis $\partial_1, \ldots, \partial_n$ of the tangent space. But all k-forms arise from linear combinations of the wedge products of the dual basis elements (see (5) in Section 8.2), and this yields the following corollary.

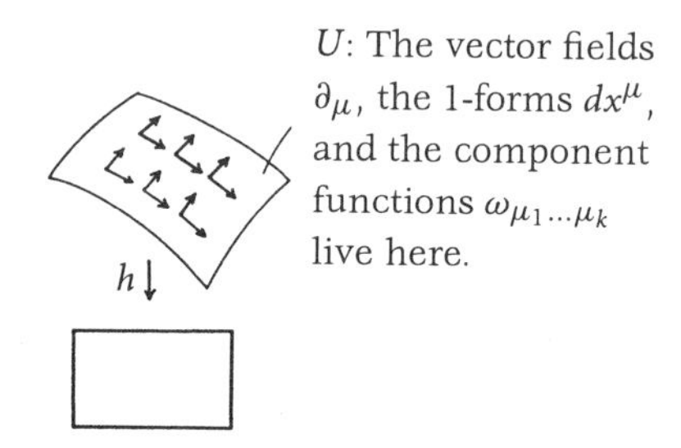

Figure 8.1. A reminder

Corollary. *If* $\omega \in \Omega^k M$ *and* (U, h) *is a chart, then*

$$\omega|U = \sum_{\mu_1 < \cdots < \mu_k} \omega_{\mu_1 \ldots \mu_k} dx^{\mu_1} \wedge \ldots \wedge dx^{\mu_k},$$

where $\omega_{\mu_1 \ldots \mu_k} := \omega(\partial_{\mu_1}, \ldots, \partial_{\mu_k}) : U \to \mathbb{R}$ *are the component functions of* ω *with respect to* (U, h).

Defining theorem (Cartan derivative). *If M is a manifold, then there is exactly one way to introduce a sequence of linear maps*

$$0 \to \Omega^0 M \xrightarrow{d} \Omega^1 M \xrightarrow{d} \Omega^2 M \xrightarrow{d} \cdots$$

so that the following three conditions are satisfied:

(a) Differential condition: *For* $f \in \Omega^0 M$, df *has its usual meaning as the differential of* f.

(b) Complex property: $d \circ d = 0$.

(c) Product rule: $d(\omega\wedge\eta) = d\omega\wedge\eta+(-1)^r\omega\wedge d\eta$ *for* $\omega \in \Omega^r M$.

We call $d\omega$ the ***exterior***, or ***Cartan, derivative*** of the differential form ω, and the entire sequence the ***de Rham complex*** of M.

The proof of the theorem is carried out in two steps, to which the next two sections are devoted.

8.4 Proof for a Chart Domain

Throughout the proof we will use the notation d_M for the Cartan derivative to be constructed and reserve d for the ordinary differential of functions. We begin by proving the theorem just for a chart domain rather than for M. There is an obvious starting point: If (U, h) is a chart, then as we saw earlier, any $\omega \in \Omega^k U$ can be written as

$$\omega = \sum_{\mu_1<\cdots<\mu_k} \omega_{\mu_1\ldots\mu_k} dx^{\mu_1} \wedge \ldots \wedge dx^{\mu_k}.$$

Using the *wedge product*, we can thus express ω in terms of *functions* and *differentials*, and these concepts are exactly what conditions (a)–(c) are about. We take the formula

$$(*) \qquad d_U\omega = \sum_{\mu_1<\cdots<\mu_k} d\omega_{\mu_1\ldots\mu_k} \wedge dx^{\mu_1} \wedge \ldots \wedge dx^{\mu_k}$$

as an established fact for the proof of uniqueness, and as a definition for the proof of existence. More precisely: If the d_U have properties (a), (b), and (c) for $M := U$, then formula $(*)$ obviously follows for all $\omega \in \Omega^k U$, and this proves the uniqueness statement for the case $M = U$. For the proof of existence, we now use $(*)$ as our definition. The maps

$$0 \to \Omega^0 U \xrightarrow{d_U} \Omega^1 U \xrightarrow{d_U} \cdots$$

defined in this way are clearly linear, and the differential condition (a) is satisfied. The complex property and the product rule still have to be verified. We begin with the product rule. Without loss of generality, let

$$\omega = f\, dx^{\mu_1} \wedge \ldots \wedge dx^{\mu_r} \quad \text{and} \quad \eta = g\, dx^{\nu_1} \wedge \ldots \wedge dx^{\nu_s}.$$

Then, by the definition (*),

$$d_U(\omega \wedge \eta) = d(fg) \wedge dx^{\mu_1} \wedge \ldots \wedge dx^{\mu_r} \wedge dx^{\nu_1} \wedge \ldots \wedge dx^{\nu_s}.$$

It follows from the usual product rule $d(fg) = df \cdot g + f \cdot dg$ for functions and the anticommutativity of the wedge product that

$$\begin{aligned} d_U(\omega \wedge \eta) & \\ &= (df \wedge dx^{\mu_1} \wedge \ldots \wedge dx^{\mu_r}) \wedge (g\, dx^{\nu_1} \wedge \ldots \wedge dx^{\nu_s}) \\ &\quad + (-1)^r (f\, dx^{\mu_1} \wedge \ldots \wedge dx^{\mu_r}) \wedge (dg \wedge dx^{\nu_1} \wedge \ldots \wedge dx^{\nu_s}) \\ &= (d_U\omega) \wedge \eta + (-1)^r \omega \wedge d_U\eta, \end{aligned}$$

as was to be proved.

Now for the complex property. We have to show that $d_U d_U \omega = 0$ for all $\omega \in \Omega^k U$. By the defining formula (*), $d_U\omega$ is a sum of wedge products of differentials, so by the product rule, which has already been proved, it suffices to consider the case $k = 0$. But for a function $f \in \Omega^0 U$ we have

$$\begin{aligned} d_U d_U f = d_U df &= d_U \sum_{\mu=1}^{n} \partial_\mu f \cdot dx^\mu \\ &= \sum_{\mu=1}^{n} d(\partial_\mu f) \wedge dx^\mu \\ &= \sum_{\mu,\nu=1}^{n} \partial_\mu \partial_\nu f \cdot dx^\nu \wedge dx^\mu = 0, \end{aligned}$$

because $\partial_\nu \partial_\mu f$ is symmetric in μ and ν and $dx^\nu \wedge dx^\mu$ is skew-symmetric. Thus we have also proved the complex property for d_U, and this completes the proof of the theorem for the special case $M = U$.

8.5 Proof for the Whole Manifold

We now turn to the general case. For the proof of existence, we will try to define d_M locally by means of charts. For

$\omega \in \Omega^k M$, we set

$$(d_M\omega)_p := (d_U\omega)_p$$

for a chart (U, h) around p, where $d_U\omega$ of course means $d_U(\omega|U)$. This is clearly independent of the choice of chart, since

$$d_U\omega|U \cap V = d_{U\cap V}\omega = d_V\omega|U \cap V$$

follows immediately from the defining formula $(*)$ for the Cartan derivative in chart domains. The d_M defined in this way obviously has the property $d_M\omega|U = d_U\omega$, so it really does define linear maps

$$0 \to \Omega^0 M \xrightarrow{d_M} \Omega^1 M \xrightarrow{d_M} \cdots$$

that satisfy the differential condition, the complex property, and the product rule. This completes the proof of existence.

For the proof of uniqueness, we must now show conversely that if d_M is a Cartan derivative for all of M, i.e. if it satisfies conditions (a)–(c), then $(d_M\omega)_p = (d_U\omega)_p$. Now,

$$\omega|U = \sum_{\mu_1<\cdots<\mu_k} \omega_{\mu_1\ldots\mu_k} dx^{\mu_1} \wedge \ldots \wedge dx^{\mu_k}.$$

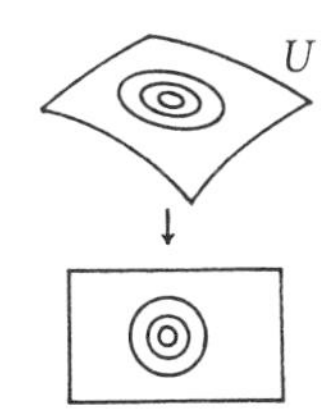

Figure 8.2. Preparation for the mesa function

But we cannot make direct use of this equation to evaluate $d_M\omega$. By hypothesis, d_M acts only on differential forms that are defined everywhere on M, and the functions $\omega_{\mu_1\ldots\mu_k}$ and the 1-forms dx^{μ_i} are just defined in U. So we resort to a trick. In $h(U)$ we choose three concentric open balls about $h(p)$, with radii $0 < \varepsilon_1 < \varepsilon_2 < \varepsilon_3$. Let their preimages under h be $U_1 \subset U_2 \subset U_3$. Now we choose a C^∞ function $\tau : U_3 \to [0, 1]$ with $\tau|U_1 \equiv 1$ and $\tau|U_3 \setminus U_2 \equiv 0$, a "mesa function," so to speak, with a plateau over U_1 and a slope in $U_2 \setminus U_1$. To do this, all we need is a C^∞ auxiliary function $\lambda : \mathbb{R}_+ \to [0, 1]$ as in Figure 8.3, with which we then define $\tau(q) := \lambda(\|h(q)\|)$ for $q \in U_3 \subset U$. The purpose of this device τ is to extend the functions $\omega_{\mu_1\ldots\mu_k}$ and $x^1, \ldots, x^n$ differentiably from U_1 to all

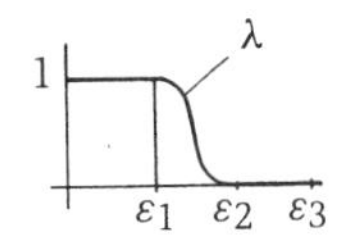

Figure 8.3. Auxiliary function for the mesa function

of M. We do this simply by defining

$$a_{\mu_1\ldots\mu_k}(q) := \begin{cases} \tau(q)\omega_{\mu_1\ldots\mu_k}(q) & \text{for } q \in U_3, \\ 0 & \text{for } q \in M \setminus U_3, \end{cases}$$

$$\xi^\mu(q) := \begin{cases} \tau(q)x^\mu(q) & \text{for } q \in U_3, \\ 0 & \text{for } q \in M \setminus U_3. \end{cases}$$

For the k-form $\widetilde{\omega} \in \Omega^k M$ given by

$$\widetilde{\omega} := \sum_{\mu_1<\cdots<\mu_k} a_{\mu_1\ldots\mu_k} d\xi^{\mu_1} \wedge \ldots \wedge d\xi^{\mu_k},$$

axioms (a)–(c) actually do imply that

$$d_M\widetilde{\omega} := \sum_{\mu_1<\cdots<\mu_k} da_{\mu_1\ldots\mu_k} \wedge d\xi^{\mu_1} \wedge \ldots \wedge d\xi^{\mu_k}$$

because the a's and the ξ's are now differentiable on all of M. In particular, as the defining formula $(*)$ for U shows,

$$(d_M\widetilde{\omega})_p = (d_U\widetilde{\omega})_p,$$

and the right-hand side is equal to $(d_U\omega)_p$ because ω and $\widetilde{\omega}$ agree on the neighborhood $U_1 \subset U$ of p. The only thing left to prove is that

$$(d_M\widetilde{\omega})_p = (d_M\omega)_p.$$

In much the same way as we chose the mesa function τ earlier, we now choose a C^∞ function $\sigma : M \to [0, 1]$ with $\sigma|M \setminus U_1 \equiv 1$ and $\sigma(p) = 0$. Then

$$\widetilde{\omega} - \omega = \sigma \cdot (\widetilde{\omega} - \omega),$$

and it follows from (a) – (c) for d_M that

$$d_M(\widetilde{\omega} - \omega) = d\sigma \wedge (\widetilde{\omega} - \omega) + \sigma d_M(\widetilde{\omega} - \omega).$$

Both the summands vanish at p because $\widetilde{\omega} - \omega$ and σ are zero there. Hence $d_M(\widetilde{\omega} - \omega)_p = 0$, as was to be shown. □

8.6 The Naturality of the Cartan Derivative

The Cartan derivative is now at our disposal. The general properties (a)–(c) enabled us to characterize it, and the local formula

$$d\omega|U = \sum_{\mu_1<\cdots<\mu_k} d\omega_{\mu_1\ldots\mu_k} \wedge dx^{\mu_1} \wedge \ldots \wedge dx^{\mu_k}$$

we obtained along the way gives us concrete instructions for computing $d\omega$ in the coordinates of a chart (U, h). There was no need to include the *naturality* of the Cartan derivative among the characterizing conditions; it now follows automatically.

Lemma. *The Cartan derivative is compatible with differentiable maps. In other words, if $f : M \to N$ is a differentiable map, then*

$$f^*d\omega = d(f^*\omega)$$

for all differential forms ω on N.

PROOF. For 0-forms $\omega \in \Omega^0 N$ (differentiable functions $\omega : N \to \mathbb{R}$), the statement $f^*d\omega = d(f^*\omega)$ is just another way of writing the chain rule since $f^*\omega := \omega \circ f$ and $(f^*d\omega)_p := d\omega_{f(p)} \circ df_p$. For differential forms of higher degree, we know in advance from the formula above for computing $d\omega|U$ that the Cartan derivative is compatible with inclusions of open sets, so we may assume without loss of generality that there is a chart (U, h) on N whose chart domain is all of N. We then have

$$\omega = \sum_{\mu_1<\cdots<\mu_k} \omega_{\mu_1\ldots\mu_k} dx^{\mu_1} \wedge \ldots \wedge dx^{\mu_k} \quad \text{and}$$

$$d\omega = \sum_{\mu_1<\cdots<\mu_k} d\omega_{\mu_1\ldots\mu_k} \wedge dx^{\mu_1} \wedge \ldots \wedge dx^{\mu_k}.$$

Applying f^* gives

$$f^*\omega = \sum_{\mu_1<\cdots<\mu_k} f^*\omega_{\mu_1\ldots\mu_k} \cdot f^*dx^{\mu_1} \wedge \ldots \wedge f^*dx^{\mu_k} \quad \text{and}$$

$$f^*d\omega = \sum_{\mu_1<\cdots<\mu_k} f^*d\omega_{\mu_1\ldots\mu_k} \wedge f^*dx^{\mu_1} \wedge \ldots \wedge f^*dx^{\mu_k}.$$

Now, before applying d to the first of these two equations so as to compare $df^*\omega$ with $f^*d\omega$, we want to convince ourselves that

$$d(f^*dx^{\mu_1} \wedge \ldots \wedge f^*dx^{\mu_k}) = 0.$$

But this follows by induction and the product rule from the known equality

$$f^*dx^{\mu_i} = d(f^*x^{\mu_i})$$

for the 0-form x^{μ_i} on N. Hence $d(f^*dx^{\mu_i}) = 0$ because $dd = 0$. Thus applying d to $f^*\omega$ gives only

$$df^*\omega = \sum df^*\omega_{\mu_1\ldots\mu_k} \wedge f^*dx^{\mu_1} \wedge \ldots \wedge f^*dx^{\mu_k}$$

by the product rule. Since d and f^* commute when applied to the 0-form $\omega_{\mu_1\ldots\mu_k}$, it follows that $df^*\omega = f^*d\omega$.

8.7 The de Rham Complex

The naturality of d also means that every differentiable map $f : M \to N$ induces a ***chain map*** between the de Rham complexes of N and M. In other words, the diagram

$$\begin{array}{ccccccccc}
0 & \longrightarrow & \Omega^0 N & \xrightarrow{d} & \Omega^1 N & \xrightarrow{d} & \Omega^2 N & \longrightarrow & \cdots \\
 & & \downarrow f^* & & \downarrow f^* & & \downarrow f^* & & \\
0 & \longrightarrow & \Omega^0 M & \xrightarrow[d]{} & \Omega^1 M & \xrightarrow[d]{} & \Omega^2 M & \longrightarrow & \cdots
\end{array}$$

commutes. The de Rham complex canonically defines a contravariant functor from the differentiable category to the category of complexes and their chain maps, as announced in Section 7.5.

Of course, the de Rham complex of an n-dimensional manifold M is interesting only up to degree n since $\Omega^k M = 0$ for $k > n$. For this reason, the *finite* sequence

$$0 \longrightarrow \Omega^0 M \xrightarrow{d} \Omega^1 M \xrightarrow{d} \cdots \xrightarrow{d} \Omega^{n-1} M \xrightarrow{d} \Omega^n M \xrightarrow{d} 0$$

is also often called the de Rham complex of M. But the naturality of d refers to more than maps between manifolds of the same dimension, so it is formally more convenient to extend the de Rham complex on the right with zeros. If $\dim N =: k < n$, then the naturality of d makes a further nontrivial statement about k-forms on M: All the k-forms coming from N have Cartan derivative zero, or are said to be "cocycles":

$$\begin{array}{ccc} \Omega^k N & \longrightarrow & 0 \\ {\scriptstyle f^*}\big\downarrow & & \big\downarrow \\ \Omega^k M & \xrightarrow{d} & \Omega^{k+1} M \end{array}$$

Corollary. *If M is an n-dimensional manifold-with-boundary and $f : M \to \partial M$ is any differentiable map, then*

$$df^*\omega = 0$$

for all $\omega \in \Omega^{n-1}\partial M$.

8.8 Test

(1) Let $(e_1, \ldots, e_n)$ be a basis of V and $(\delta^1, \ldots, \delta^n)$ its dual basis. Then the following family of wedge products is a basis of $\mathrm{Alt}^2 V$:

- ☐ $(\delta^\mu \wedge \delta^\nu)_{\mu,\nu=1,\ldots,n}$.
- ☐ $(\delta^\mu \wedge \delta^\nu)_{\mu \le \nu}$.
- ☐ $(\delta^\mu \wedge \delta^\nu)_{\mu < \nu}$.

(2) Let V be an n-dimensional vector space. Which of the following conditions on k, with $0 \le k \le n$, is equivalent to $\omega \wedge \omega = 0$ for all $\omega \in \mathrm{Alt}^k V$?

☐ $0 < k$.

☐ $2k > n$.

☐ k is odd or $2k > n$.

(3) Let V be as above and $0 \leq r \leq n$. Without further conditions on r, does $\eta \mapsto \ldots \wedge \eta$ give an isomorphism

$$\text{Alt}^{n-r} V \xrightarrow{\cong} \text{Hom}(\text{Alt}^r V, \text{Alt}^n V)?$$

☐ Yes. The spaces have the same dimension and the homomorphism is clearly injective.

☐ No. For example, the homomorphism is *not* injective for odd r and $n = 2r$ because $\eta \wedge \eta = 0$.

☐ Only if r and $n - r$ are both even.

(4) In local coordinates (x, y) on a two-dimensional manifold,

☐ $dx \wedge dy\, (\partial_y, \partial_x) = 1$.

☐ $dx \wedge dy\, (\partial_y, \partial_x) = 0$.

☐ $dx \wedge dy\, (\partial_y, \partial_x) = -1$.

(5) In local coordinates (t, x, y, z) on a four-dimensional manifold,

☐ $dt \wedge dx\, (\partial_y, \partial_z) = 1$.

☐ $dt \wedge dx\, (\partial_y, \partial_z) = 0$.

☐ $dt \wedge dx\, (\partial_y, \partial_z) = -1$.

(6) Let $\omega \in \Omega^r M$, $\eta \in \Omega^s M$, and $\zeta \in \Omega^t M$. Then the signs in the formula

$$d(\omega \wedge \eta \wedge \zeta) = \pm d\omega \wedge \eta \wedge \zeta \pm \omega \wedge d\eta \wedge \zeta \pm \omega \wedge \eta \wedge d\zeta$$

are, in order,

☐ $+1, \ +1, \ +1$.

☐ $(-1)^r, \ (-1)^s, \ (-1)^t$.

☐ $+1, \ (-1)^r, \ (-1)^{r+s}$.

(7) For the coordinate functions x and y on $\mathbb{R}^2$, we have

- ☐ $d(x\,dy + y\,dx) = 0$.
- ☐ $d(x\,dx + y\,dy) = 0$.
- ☐ $d(xy\,dx + yx\,dy) = 0$.

(8) Let $f : M \to N$ be a differentiable map between manifolds. Is the composition of the three homomorphisms

$$\Omega^{r-1}N \xrightarrow{d} \Omega^r N \xrightarrow{f^*} \Omega^r M \xrightarrow{d} \Omega^{r+1}M$$

necessarily zero?

- ☐ Yes, because of the naturality of the Cartan derivative.
- ☐ No. Taking $M = N = \mathbb{R}^2$, $f(x, y) := (y, x)$, and $\omega = xy \in \Omega^0 N$ gives a counterexample: $d\omega = (dx)y - x\,dy = y\,dx - x\,dy$, so $f^*d\omega = x\,dy - y\,dx$ and hence $df^*d\omega = dx \wedge dy - dy \wedge dx = 2\,dx \wedge dy \neq 0$.
- ☐ No. If we take $N = M = \mathbb{R}$, for instance, and set $\omega = f$, we get $df^*df = \|df\|^2$. This doesn't vanish in general.

(9) Let r and φ denote the usual polar coordinates in the plane. Then $r\,dr \wedge d\varphi =$

- ☐ $dx \wedge dy$.
- ☐ $dy \wedge dx$.
- ☐ $\sqrt{x^2 + y^2}\,dx \wedge dy$.

(10) On a manifold M, one can also consider *complex-valued* differential forms $\omega \in \Omega^r(M, \mathbb{C})$, extend the wedge product of real forms to a complex-bilinear operation on complex-valued forms, and (by splitting into real and imaginary parts) define the Cartan derivative for complex-valued forms as well. Which of the following is true on $M := \mathbb{C}$?

- ☐ $dz \wedge d\bar{z} = dx \wedge dy$.

☐ $dz \wedge d\bar{z} = 2\,dx \wedge dy$.

☐ $dz \wedge d\bar{z} = -2i\,dx \wedge dy$.

8.9 Exercises

EXERCISE 8.1. Let V be an n-dimensional real vector space. Show that the alternating k-linear map $V^* \times \cdots \times V^* \xrightarrow{u} \mathrm{Alt}^k V$ defined by $(\varphi^1, \ldots, \varphi^k) \mapsto \varphi^1 \wedge \ldots \wedge \varphi^k$ is universal in the following sense: For every alternating k-linear map $\alpha : V^* \times \cdots \times V^* \to W$ there is exactly one linear map $f : \mathrm{Alt}^k V \to W$ with $\alpha = f \circ u$.

EXERCISE 8.2. Consider the usual coordinate functions x, y, and z on $\mathbb{R}^3$. Give a 2-form $\omega \in \Omega^2\mathbb{R}^3$ such that

$$d\omega = dx \wedge dy \wedge dz.$$

Does $\omega = d\eta$ for some $\eta \in \Omega^1\mathbb{R}^3$?

EXERCISE 8.3. Let M be an n-dimensional manifold and let $\omega \in \Omega^{n-1}M$. Show that in local coordinates

$$d\omega(\partial_1, \ldots, \partial_n) = \sum_{\mu=1}^{n} (-1)^{\mu-1} \partial_\mu \omega_{1\ldots\hat{\mu}\ldots n}.$$

EXERCISE 8.4. Let $\omega := dx^1 \wedge \ldots \wedge dx^n \in \Omega^n\mathbb{R}^n$ and let $v = v^\mu \partial_\mu$ be a vector field on $\mathbb{R}^n$. Determine $\eta := v \lrcorner\, \omega \in \Omega^{n-1}\mathbb{R}^n$ and $d\eta \in \Omega^n\mathbb{R}^n$. Also give an explicit vector field v such that, on S^{n-1}, η induces the canonical volume form of S^{n-1}.

8.10 Hints for the Exercises

FOR EXERCISE 8.1. For this purely linear-algebraic exercise, you have to recall the linear-algebraic fact that given a basis $a_1, \ldots, a_m$ of a vector space A and elements $b_1, \ldots, b_m$ of a vector space B, there is exactly one linear map $f : A \to B$ such that $f(a_i) = b_i$ for $i = 1, \ldots, m$.

FOR EXERCISE 8.2. Here the coordinates x, y, z define a chart (U, h) with $U = M = \mathbb{R}^3$, so the local defining formula for the Cartan derivative (emphasized again at the beginning of Section 8.6) holds immediately for $\omega|U = \omega$ itself. It's just that x^1, x^2, x^3 are renamed x, y, z.

FOR EXERCISE 8.3. This is another direct application of the local formula for the Cartan derivative (see Section 8.6).

FOR EXERCISE 8.4. We were introduced to the notation $v \lrcorner \omega$ for "v in ω" in Exercise 3.1—there in the linear-algebraic situation $v \in V$, $\omega \in \text{Alt}^n V$, here to be applied analogously to a vector field v on M and a form $\omega \in \Omega^n M$. The "canonical volume form" $\omega_{S^{n-1}} \in \Omega^{n-1} S^{n-1}$ is the $(n-1)$-form that responds with $+1$ to every positively oriented orthonormal basis $(v_1, \ldots, v_{n-1})$ of $T_p S^{n-1}$. "Positively oriented" means that we think of S^{n-1} as ∂D^n, with the induced orientation. (Note: If $(V, \langle \cdot, \cdot \rangle)$ is an oriented n-dimensional Euclidean vector space and $(v_1, \ldots, v_n)$, $(v_1', \ldots, v_n')$ are positively oriented orthonormal bases, then the automorphism $f : V \to V$ with $v_i \mapsto v_i'$ has determinant $+1$. Hence $\omega_{S^{n-1}}$ is well defined; see Exercise 3.2.)

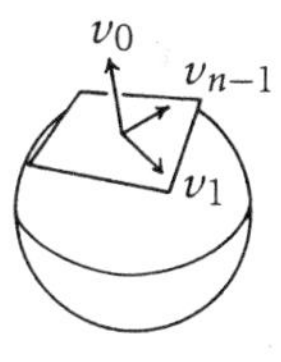

Figure 8.4.

Here's the real hint for Exercise 8.4: $\mathbb{R}^n$ also has a canonical volume form, namely

$$\omega = dx^1 \wedge \ldots \wedge dx^n.$$

How does $dx^1 \wedge \ldots \wedge dx^n$ respond to $(v_0, \ldots, v_{n-1})$, and what does this have to do with $v \lrcorner \omega$?

9 CHAPTER Stokes's Theorem

9.1 The Theorem

Now we finally come to the theorem so much of our discussion has been about.

Stokes's Theorem. *Let M be an oriented n-dimensional manifold-with-boundary and $\omega \in \Omega^{n-1}M$ an $(n-1)$-form with compact support. Then*

$$\int_M d\omega = \int_{\partial M} \omega.$$

Before we begin the proof, we recall two conventions that were used implicitly in formulating the theorem. First, ∂M is oriented according to the orientation convention established in Section 6.8: the outward normal followed by the orientation of the boundary gives the orientation of M. Second, $\int_{\partial M} \omega := \int_{\partial M} \iota^* \omega$, where $\iota : \partial M \hookrightarrow M$ is the inclusion map (see Section 7.2 for this notation). We carry out the proof in three steps of increasing generality:

1. The case $M = \mathbb{R}^n_-$.
2. The case where there exists a chart (U, h) with $\operatorname{supp} \omega \subset U$.
3. The general case.

The first step takes some computation but no ideas; a completely straightforward application of the definitions serves our purpose. The other two are more abstract. The third and last step introduces *partitions of unity*, a tool that is often useful elsewhere in passing from local to global situations.

9.2 Proof for the Half-Space

Let $M = \mathbb{R}^n_-$. In the canonical coordinates,

$$\omega = \sum_{\mu=1}^{n} \omega_{1\ldots\widehat{\mu}\ldots n} dx^1 \wedge \ldots \widehat{\mu} \ldots \wedge dx^n,$$

or, if we use the abbreviation $f_\mu := \omega_{1\ldots\widehat{\mu}\ldots n}$ for the component functions,

$$\omega = \sum_{\mu=1}^{n} f_\mu dx^1 \wedge \ldots \widehat{\mu} \ldots \wedge dx^n,$$

where the notation $\widehat{\mu}$ again means that the index μ or the corresponding factor dx^μ is to be omitted.

The two integrands $d\omega \in \Omega^n \mathbb{R}^n_-$ and $\iota^*\omega \in \Omega^{n-1}\mathbb{R}^{n-1}$ can be computed from this formula and the definitions as follows:

$$\begin{aligned} d\omega &= \sum_{\mu=1}^{n} df_\mu \wedge dx^1 \wedge \ldots \widehat{\mu} \ldots \wedge dx^n \\ &= \sum_{\mu=1}^{n} \left(\sum_{\nu=1}^{n} \partial_\nu f_\mu dx^\nu \right) \wedge dx^1 \ldots \widehat{\mu} \ldots \wedge dx^n \\ &= \sum_{\mu=1}^{n} (-1)^{\mu-1} \partial_\mu f_\mu \cdot dx^1 \wedge \ldots \wedge dx^n. \end{aligned}$$

If we also denote the canonical coordinates of $\{0\} \times \mathbb{R}^{n-1} \subset \mathbb{R}^n$ by $x^2, \dots, x^n$, then

$$\begin{aligned}\iota^*\omega &= \sum_{\mu=1}^{n} \iota^* f_\mu \cdot \iota^* dx^1 \wedge \dots \widehat{\mu} \dots \wedge dx^n \\ &= \iota^* f_1 \cdot dx^2 \wedge \dots \wedge dx^n \in \Omega^{n-1}\mathbb{R}^{n-1},\end{aligned}$$

since the inclusion $\iota : \{0\} \times \mathbb{R}^{n-1} \hookrightarrow \mathbb{R}^n$ obviously induces the functions $0, x^2, \dots, x^n$ on $\{0\} \times \mathbb{R}^{n-1}$ from the coordinate functions $x^1, \dots, x^n$ on $\mathbb{R}^n$. Hence

$$\begin{aligned}\iota^* dx^1 &= 0 && \text{and} \\ \iota^* dx^\mu &= dx^\mu && \text{for } \mu \geq 2.\end{aligned}$$

So much for the integrands $d\omega$ on $\mathbb{R}^n_-$ and $\iota^*\omega$ on $\partial\mathbb{R}^n_-$. We now turn to the integral itself. The canonical coordinates on $\mathbb{R}^n_-$ define an orientation-preserving chart, of course, and according to the orientation convention so do the coordinates $x^2, \dots, x^n$ of $\partial\mathbb{R}^n_-$. Hence, by the definition of the integral (integration of the "downstairs component function"),

$$\int_{\mathbb{R}^n_-} d\omega = \sum_{\mu=1}^{n} \int_{\mathbb{R}^n_-} (-1)^{\mu-1} \partial_\mu f_\mu dx^1 \dots dx^n \quad \text{and}$$

$$\int_{\partial\mathbb{R}^n_-} \omega = \int_{\mathbb{R}^{n-1}} f_1(0, x^2, \dots, x^n) dx^2 \dots dx^n$$

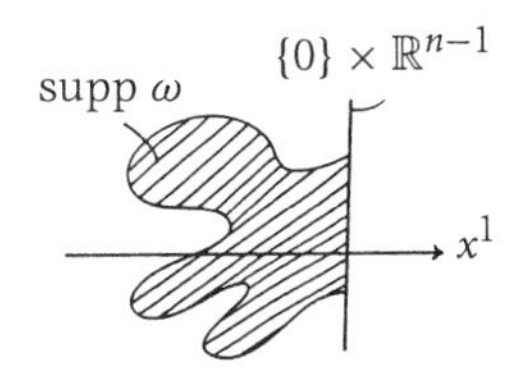

Figure 9.1. Stokes's theorem in the case $M = \mathbb{R}^n_-$

as ordinary multiple integrals of differentiable integrands with compact support. By Fubini's theorem, we may integrate with respect to the individual variables in any order. Thus in the μth summand of $\int_M d\omega$ we may integrate first with respect to the μth variable. Since the support $\overline{\{x \in \mathbb{R}^n_- : \omega_x \neq 0\}}$ of ω is bounded, so is the support of f, and we obtain

$$\int_{-\infty}^{0} \partial_1 f_1 dx^1 = \Big[f_1\Big]_{x^1=-\infty}^{x^1=0} = f_1(0, x^2, \dots, x^n)$$

for $\mu = 1$ but

$$\int_{-\infty}^{\infty} \partial_\mu f_\mu dx^\mu = \Big[f_\mu\Big]_{x^\mu=-\infty}^{x^\mu=+\infty} \equiv 0$$

for the other μ's. Hence

$$\int_M d\omega = \int_{\mathbb{R}^{n-1}} f_1(0, x^2, \dots, x^n) dx^2 \dots dx^n = \int_{\partial M} \omega$$

for our first case $M := \mathbb{R}^n_-$.

9.3 Proof for a Chart Domain

Let (U, h) be a chart on M with $\operatorname{supp}\omega \subset U$. Our definition of manifolds-with-boundary allows the two possibilities that $h(U)$ is open in $\mathbb{R}^n_-$ or in $\mathbb{R}^n$. Without loss of generality we may assume the former here, since by the compactness of $\operatorname{supp}\omega$ we could always achieve it if necessary by translating and shrinking the chart domain. We may further assume that $h : U \to U'$ preserves orientation and hence, by the orientation convention, that so does $h|\partial U : \partial U \to \partial U'$. But then

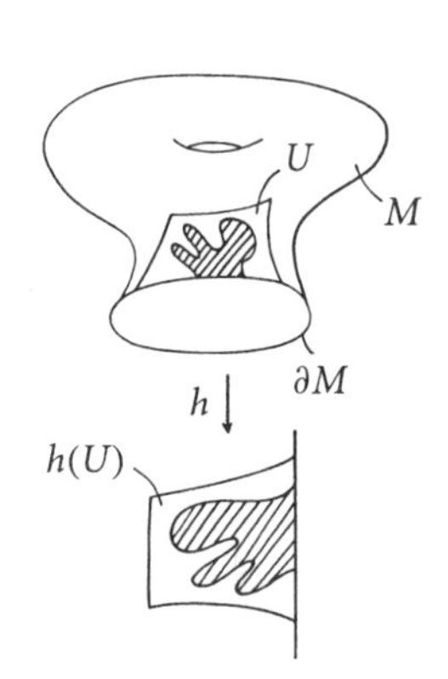

Figure 9.2. Stokes's theorem in the case $\operatorname{supp}\omega \subset U$

$$\int_M d\omega = \int_U d\omega = \int_{h(U)} h^{-1*} d\omega = \int_{h(U)} d(h^{-1*}\omega),$$

by the change-of-variables formula for integration on manifolds (see Section 5.5) and the naturality of the Cartan derivative. We now extend $h^{-1*}\omega$ to a form $\omega' \in \Omega^{n-1}\mathbb{R}^n_-$ by setting it equal to zero outside $h(U)$, which is possible because $\operatorname{supp} h^{-1*}\omega = h(\operatorname{supp}\omega)$ is compact. Then

$$\int_{h(U)} d(h^{-1*}\omega) = \int_{\mathbb{R}^n_-} d\omega' \underset{\text{Case 1}}{=} \int_{\partial\mathbb{R}^n_-} \omega' = \int_{h(\partial U)} h^{-1*}\omega.$$

But the change-of-variables formula for $h|\partial U : \partial U \to \partial U'$ gives

$$\int_{h(\partial U)} h^{-1*}\omega = \int_{\partial U} \omega = \int_{\partial M} \omega,$$

and this completes the second step.

9.4 The General Case

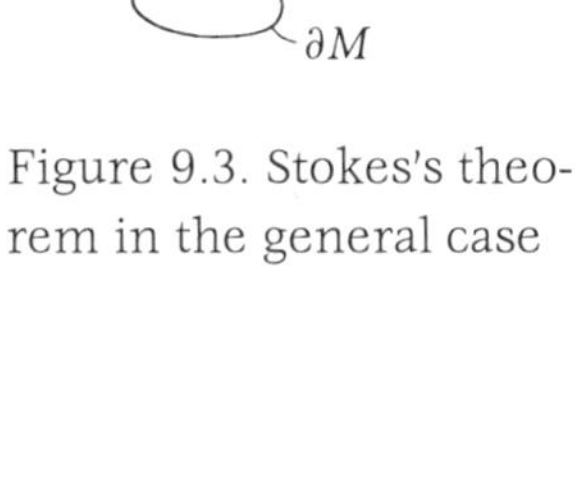

Figure 9.3. Stokes's theorem in the general case

Everything so far has been routine, but now we need a trick. The support may no longer fit inside a chart domain, and decomposing M or $\operatorname{supp}\omega$ by brute force into small measurable pieces would lead to discontinuous integrands in $\mathbb{R}^n_-$ to which the Cartan derivative could not be applied at all. If only we could write ω as a sum $\omega = \omega_1 + \cdots + \omega_r$ of *differentiable* $(n-1)$-forms $\omega_i \in \Omega^{n-1}M$, each of which had compact support $\operatorname{supp}\omega_i$ that fit inside a chart domain U_i! By Section 9.3, this would certainly complete the proof.

And this is exactly what we'll manage to do now. First, around each $p \in \operatorname{supp}\omega$ we choose an orientation-preserving chart (U_p, h_p) and a C^∞ function $\lambda_p : M \to [0,1]$ such that $\lambda_p(p) > 0$ and the support of λ_p is compact and contained in U_p. That's no problem: we need only lift a suitable "bump function" β_p with compact support in $h(U_p)$ up to U_p by setting $\lambda_p(q) := \beta_p(h(q))$ for $q \in U_p$ and 0 otherwise. Then $\{\lambda_p^{-1}(0,1]\}_{p\in\operatorname{supp}\omega}$ is a family of open sets whose union contains $\operatorname{supp}\omega$, and since $\operatorname{supp}\omega$ is compact there are finitely many $p_1, \ldots, p_r$ such that

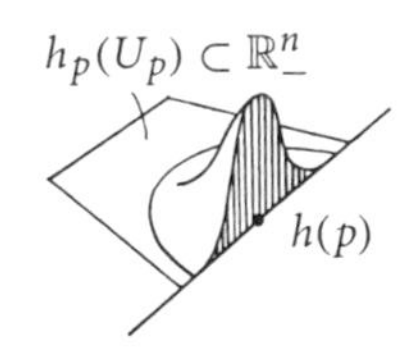

Figure 9.4. β_p for $p \in \partial M$

$$\operatorname{supp}\omega \subset \bigcup_{i=1}^{r} \lambda_{p_i}^{-1}(0,1] =: X.$$

On the open set $X \subset M$ we now define r differentiable functions $\tau_1, \ldots, \tau_r$ by

$$\begin{aligned} \tau_i : \quad X &\longrightarrow [0,1], \\ x &\mapsto \frac{\lambda_{p_i}(x)}{\lambda_{p_1}(x) + \cdots + \lambda_{p_r}(x)}. \end{aligned}$$

Then obviously

$$\sum_{i=1}^{r} \tau_i(x) = 1 \quad \text{for all } x \in X,$$

which is why $\{\tau_i\}_{i=1,\ldots,r}$ is also called a "partition of unity" on X. Multiplication by ω now gives us our corresponding "partition of ω." More precisely, we define $\omega_i \in \Omega^{n-1}M$ by

$$\omega_{ip} := \begin{cases} \tau_i(p)\omega_p & \text{for } p \in X, \\ 0 & \text{otherwise.} \end{cases}$$

Since $\operatorname{supp}\omega$ is compact, so is $\operatorname{supp}(\tau_i \cdot \omega|X) \subset \operatorname{supp}\omega$; hence ω_i is differentiable not only on X but on all of M, and it follows from $\operatorname{supp}\omega \subset X$ and $\sum \tau_i \equiv 1$ on X that

$$\omega = \omega_1 + \cdots + \omega_r.$$

Finally, the supports of the individual summands fit as desired inside a chart domain, because $\omega_{ip} \neq 0$ implies that $\tau_i(p) \neq 0$ and hence that $\lambda_{p_i}(p) \neq 0$, so $\operatorname{supp}\omega_i \subset \operatorname{supp}\lambda_{p_i} \subset U_p$. □

9.5 Partitions of Unity

Stokes's theorem has now been proved. Partitions of unity are a very useful tool in other contexts as well (see, for example, [J:*Top*], Chapter VIII, §4). In particular, they give us what was promised at the end of Section 5.3, a way to define integration on manifolds without forcibly splitting the manifold into little pieces.

Definition. Let M be a manifold and $\mathfrak{U}$ an open cover of M (by the chart domains of an atlas, for instance). By a differentiable ***partition of unity*** subordinate to the cover $\mathfrak{U}$, we mean a family $\{\tau_\alpha\}_{\alpha\in A}$ of C^∞ functions $\tau_\alpha : M \to [0, 1]$ with the following three properties:

(1) The family $\{\tau_\alpha\}_{\alpha\in A}$ is locally finite, in the sense that for every $p \in M$ there is an open neighborhood V_p such that $\tau_p|V_p \equiv 0$ for all but finitely many $\alpha \in A$.

(2) $\sum_{\alpha\in A} \tau_\alpha(p) = 1$ for all $p \in M$.

(3) For every α, supp τ_α is contained in some set in the open cover $\mathfrak{U}$.

Lemma. *Every open cover of a manifold M has a subordinate partition of unity.*

PROOF. If M is compact we can proceed as in the proof of Stokes's theorem. First, for every $p \in M$ we choose a bump function $\lambda_p : M \to [0, 1]$ with support in one of the sets of the open cover and $\lambda_p(p) > 0$. We then find $p_1, \ldots, p_r$ with $\cup_{i=1}^r \lambda_{p_i}^{-1}(0, 1] = M$ and set $\tau_k := \lambda_{p_k} / \sum_{i=1}^r \lambda_{p_i}$. Of course, no problems arise with local finiteness or adding up the bump functions because there are only finitely many functions.

If M is not compact, we use a ***compact exhaustion***. This is a sequence

$$K_1 \subset K_2 \subset \cdots \subset M$$

of compact sets with $K_i \subset \overset{\circ}{K}_{i+1}$ and $\cup_{i=1}^\infty K_i = M$. Compact exhaustions are usually easy to find in concrete cases. One way to give a general existence proof is as follows: Let $\{\mathcal{O}_i\}_{i\in\mathbb{N}}$ be a countable basis for the topology of M, and without loss of generality let the closures $\overline{\mathcal{O}}_i$ all be compact. (If they are not already compact, just omit all the $\mathcal{O}_i$ with noncompact closure from the basis; the rest still form a basis.) Now, defining a sequence recursively by

$$K_i := \bigcup_{k=1}^{n_i} \overline{\mathcal{O}}_k \subset \bigcup_{k=1}^{n_{i+1}} \mathcal{O}_k, \quad \text{where } 1 = n_1 < n_2 < \cdots,$$

gives our compact exhaustion.

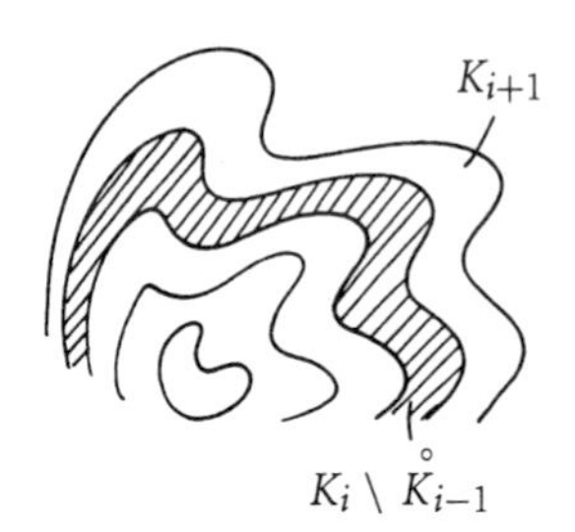

Figure 9.5. The strips $K_i \setminus \overset{\circ}{K}_{i-1}$ are "taken care of" by $\lambda_1^i, \ldots, \lambda_{r_i}^i$.

The partition of unity is obtained in the now-familiar way. For each i, finitely many differentiable (bump) functions $\lambda_1^i, \ldots, \lambda_{r_i}^i : M \to [0, 1]$ are chosen such that

$$\lambda_1^i + \cdots + \lambda_{r_i}^i > 0$$

for all x in the (compact!) set $K_i \setminus \overset{\circ}{K}_{i-1}$ but the individual supports are small enough that each fits inside *both* an element of $\mathfrak{U}$ *and* the (open!) set $\overset{\circ}{K}_{i+1} \setminus K_{i-2}$. Then the entire family $\{\lambda_j^i\}_{i\in\mathbb{N},1\le j\le r_i}$ is clearly locally finite, the function

$$\lambda := \sum_{i=1}^{\infty}\sum_{j=1}^{r_i} \lambda_j^i$$

is C^∞ and positive everywhere on M, and setting $\tau_j^i := \lambda_j^i/\lambda$ gives the desired partition of unity $\{\tau_j^i\}_{i\in\mathbb{N},1\le j\le r_i}$. □

Incidentally, *any* partition of unity $\{\tau_\alpha\}_{\alpha\in A}$ on a manifold M, not just the one we constructed in the proof of the lemma, has the property that τ_α is the zero function for all but countably (or finitely) many α's. This follows from local finiteness because manifolds satisfy the second countability axiom. So without loss of generality a partition of unity can always be thought of as $\{\tau_i\}_{i\in\mathbb{N}}$, and on compact manifolds as $\{\tau_i\}_{i=1,\ldots,r}$.

9.6 Integration via Partitions of Unity

Let M be an oriented n-dimensional manifold, and let $\{\tau_i\}_{i\in\mathbb{N}}$ be a partition of unity with each $\operatorname{supp}\tau_i$ contained in the chart domain U_i of an orientation-preserving chart (U_i, h_i). Then any n-form ω can be written as a locally finite sum $\omega = \sum_{i=1}^{\infty} \omega_i$, where

$$\omega_i := \tau_i \cdot \omega.$$

Let $a_i : h_i(U_i) \to \mathbb{R}$ denote the downstairs component function

$$a_i := \omega_{1\ldots n} \circ h_i^{-1}$$

in terms of (U_i, h_i). Then the following holds.

Note. *In this situation, ω is integrable if and only if each a_i is integrable on $h_i(U_i)$ and*

$$\sum_{i=1}^{\infty} \int_{h_i(U_i)} |a_i| \, dx < \infty.$$

Then

$$\int_M \omega = \sum_{i=1}^{\infty} \int_{h_i(U_i)} a_i \, dx.$$

Of course, this is a "note" only if integration on manifolds has already been introduced in some other way. Otherwise this formula is used as the *definition* of $\int_M \omega$, and its independence of the choice of charts and of the partition of unity need only be proved as a lemma.

How far this concept of integration on manifolds extends depends on which concept of the integral in $\mathbb{R}^n$ is used. The Lebesgue integral would give us back the concept we defined in Section 5.4. For many purposes, however, one can get by with much less: without loss of generality, the supports $\operatorname{supp} \tau_i$ of the partition of unity can always be assumed to be compact. Then for *continuous* ω, for instance, and all the more for the $\omega \in \Omega^n M$ that we always consider in the Cartan calculus and Stokes's theorem, each summand $\int_{h(U_i)} a_i dx$ is just an ordinary iterated integral

$$\int_{\alpha_n}^{\beta_n} \cdots \int_{\alpha_1}^{\beta_1} f(x_1, \ldots, x^n) \, dx^1 \ldots dx^n$$

over a *box* in $\mathbb{R}^n$ (even if $h(U_i)$ itself is unbounded). The integrand is continuous (differentiable if $\omega \in \Omega^n M$), and this situation is controlled by even the most elementary notion of integral. If in addition, as usually happens, the support of ω is assumed to be compact, then there are at most finitely many such summands, and integration on manifolds is—may one say?—quite simple. One may.

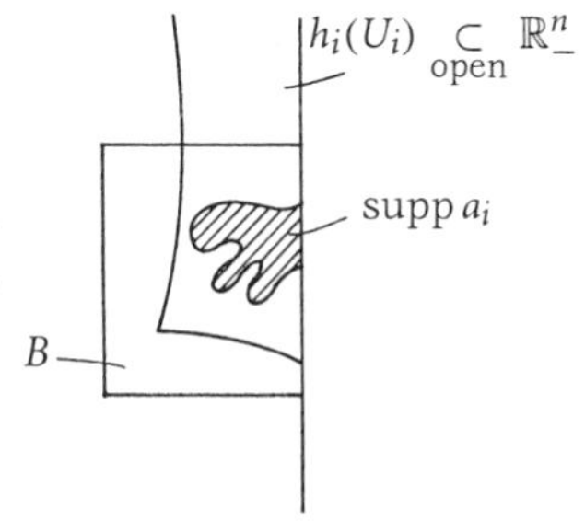

Figure 9.6. $\int_{h_i(U_i)} a_i \, dx$ is an integral of a continuous integrand over a box B in $\mathbb{R}^n$.

9.7 Test

(1) The component function of the n-form

$$dx^\mu \wedge dx^1 \wedge \ldots \widehat{\mu} \ldots \wedge dx^n$$

on $\mathbb{R}^n$ with respect to the coordinates $x^1, \ldots, x^n$ is the constant function

☐ 1. ☐ $(-1)^\mu$. ☐ $(-1)^{\mu-1}$.

(2) Let $\mu_1 < \cdots < \mu_r$ and $\omega := dx^{\mu_1} \wedge \ldots \wedge dx^{\mu_r} \in \Omega^r(\mathbb{R}^n_-)$. For the inclusion $\iota : \{0\} \times \mathbb{R}^{n-1} \hookrightarrow \mathbb{R}^n$, under what hypotheses is it true that $\iota^*\omega = 0$?

☐ When one of the μ_i equals 1.

☐ When none of the μ_i equals 1.

☐ Never.

(3) For the special case $M = \mathbb{R}^1_-$, Stokes's theorem reduces to the statement that if $f : \mathbb{R}_- \to \mathbb{R}$ is a C^∞ function with compact support, then $\int_{-\infty}^0 f'(x)dx =$

☐ 0. ☐ $f(0)$. ☐ $-f(0)$.

(4) It is clear that the hypothesis in Stokes's theorem that ω have compact support cannot just be omitted even in the case $M = \mathbb{R}^n_-$, because then the integrals may not exist. But does the theorem remain true if instead of the compactness of the support we require the existence of the integrals on both sides?

☐ Yes, because the harmless behavior of ω and $d\omega$ at infinity is an adequate substitute for compactness of the support.

☐ Yes, because this hypothesis is actually equivalent to compactness of the support.

☐ No, as you can see by looking at the case $n = 1$.

(5) On the question of whether compact subsets can be contained in chart domains: Consider

$$X := S^1 \times \{1\} \cup \{1\} \times S^1 \subset S^1 \times S^1.$$

Is X contained in a chart on the torus?

☐ No, because even $S^1 \times \{1\}$ by itself doesn't fit inside a chart domain.

☐ No, although $S^1 \times \{1\}$ and $\{1\} \times S^1$ individually fit inside chart domains. Think about the intersections of their images under a single chart that contains all of X!

☐ Yes, because the punctured torus $S^1 \times S^1 \setminus \{p\}$ is already diffeomorphic to an open subset of $\mathbb{R}^2$.

(6) Suppose that you are given the C^∞ function $f : \mathbb{R} \to \mathbb{R}$ defined by $f(x) := e^{-1/x^2}$ for $x > 0$ and $f(x) := 0$ for $x \leq 0$ and asked to construct a small bump function about the origin in $\mathbb{R}^n$, that is, a C^∞ function $\beta : \mathbb{R}^n \to \mathbb{R}_+$ with support the closed ball about 0 of radius $\varepsilon > 0$. Which of the following definitions yields the desired function?

☐ $\beta(x) := f(\varepsilon - \|x\|)$.

☐ $\beta(x) := f(\varepsilon^2 - \|x\|^2)$.

☐ $\beta(x) := f(\|x\|^2 - \varepsilon^2)$.

(7) Let $U \subset M$ be an open subset of a manifold, a chart domain for instance. Let the functions $\tau : M \to \mathbb{R}$ and $f : U \to \mathbb{R}$ be differentiable (i.e. C^∞), and let τ vanish outside U. Is the function F defined by

$$F(x) := \begin{cases} \tau(x)f(x) & \text{for } x \in U, \\ 0 & \text{for } x \in M \setminus U \end{cases}$$

differentiable on all of M?

☐ Yes, always.

☐ Yes if f is bounded. Otherwise, not in general.

☐ Boundedness is enough to make F continuous, but not differentiable.

(8) For a partition of unity $\{\tau_\alpha\}_{\alpha\in A}$ on a compact manifold M, why are there always only finitely many α with $\tau_\alpha \not\equiv 0$?

☐ Because finitely many of the open subsets

$$\{x \in M : \tau_\alpha(x) \neq 0\}$$

are already enough to cover M.

☐ Because finitely many of the sets V_p that exist by the requirement of local finiteness are already enough to cover M.

☐ Because—it isn't true at all: The supports $\operatorname{supp} \tau_\alpha$ can "keep getting smaller" even on compact manifolds, so there is room for infinitely many in a locally finite way.

(9) On an n-dimensional manifold (without boundary) M, let ω be an $(n-1)$-form with compact support and f an arbitrary differentiable function. It follows from Stokes's theorem that

☐ $\int_M f\, d\omega = 0$.

☐ $\int_M f\, d\omega = \int_M df \wedge \omega$.

☐ $\int_M f\, d\omega = -\int_M df \wedge \omega$.

(10) Let $\{\tau_\alpha\}_{\alpha\in A}$ and $\{\sigma_\lambda\}_{\lambda\in\Lambda}$ be two partitions of unity on M. Is the family $\{\tau_\alpha\sigma_\lambda\}_{(\alpha,\lambda)\in A\times\Lambda}$ also a partition of unity?

☐ Yes, always.

☐ Only if one of the two is finite (that is, if its functions vanish identically for all but finitely many indices).

☐ Only if both are finite.

9.8 Exercises

EXERCISE 9.1. Let M be an oriented compact n-dimensional manifold and (U, h) a "box-shaped" chart, i.e. one with

$$h(U) = (a_1, b_1) \times \cdots \times (a_n, b_n) \subset \mathbb{R}^n,$$

and let $\omega \in \Omega^n M$ be an n-form whose compact support is contained in U. Show directly, without using Stokes's theorem, that $\int_M d\omega = 0$.

EXERCISE 9.2. Let M be an oriented compact n-dimensional manifold and $f : M \to N$ a differentiable map to an $(n-1)$-dimensional manifold N. Also let $\eta \in \Omega^{n-1} N$ and $\omega := f^*\eta$. Show that $\int_{\partial M} \omega = 0$.

EXERCISE 9.3. Prove that on any n-dimensional orientable manifold M there is an n-form $\omega \in \Omega^n M$ with $\omega_p \neq 0$ for all $p \in M$.

EXERCISE 9.4. Let M be an n-dimensional manifold-with-boundary and $\eta \in \Omega^{n-1}\partial M$. Show that there is an $(n-1)$-form $\omega \in \Omega^{n-1} M$ with $\iota^*\omega = \eta$, where $\iota : \partial M \hookrightarrow M$ denotes the inclusion map.

9.9 Hints for the Exercises

FOR EXERCISE 9.1. This exercise is closely related to the first step in the proof of Stokes's theorem, and its only point is to give you a better understanding of that step.

FOR EXERCISE 9.2. Some exercises are so fragile that if you just *touch* them they disintegrate. So I'll keep my hands off this one and tell you instead about a nice application.

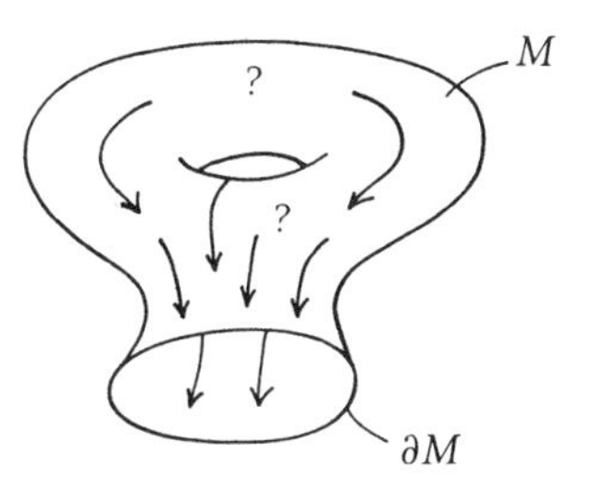

Figure 9.7.

Can an oriented manifold M be retracted differentiably onto its boundary? In other words, can you find a differentiable map $\rho : M \to \partial M$ such that the composition

$$\partial M \overset{\iota}{\hookrightarrow} M \overset{\rho}{\to} \partial M$$

is the identity? Always? Sometimes? Never? Certainly not always: a retraction $\rho : [0,1] \to \{0,1\}$ would be a continuous function with $\rho(0) = 0$ and $\rho(1) = 1$ that assumed no intermediate value. But in higher dimensions?

It hardly seems likely: the manifold will probably tear if it's retracted by force onto its boundary. Or is there a twist that could make this work after all? Perhaps in still higher dimensions?

Exercise 9.2 shows that it *never* works. To see this, choose any $\eta \in \Omega^{n-1}(\partial M)$ with $\int_{\partial M} \eta \neq 0$. This is always possible. We just have to choose a little bump function $\lambda \geq 0$ with nonempty support in a chart domain U of ∂M and set

$$\eta_p := \begin{cases} \lambda(p) dx^1 \wedge \ldots \wedge dx^{n-1} & \text{in } U, \\ 0 & \text{otherwise.} \end{cases}$$

Now, if $\rho : M \to \partial M$ were a differentiable retraction, so that $\rho \circ \iota = \mathrm{Id}_{\partial M}$, then Exercise 9.2 would give the contradiction

$$\int_{\partial M} \rho^* \eta := \int_{\partial M} \iota^* \rho^* \eta = \int_{\partial M} \eta = 0.$$

Thus we have shown:

Theorem. *No compact orientable manifold can be retracted differentiably onto its boundary.*

Corollary. *Every differentiable map $f : D^n \to D^n$ has a fixed point, because otherwise there would exist a differentiable retraction $\rho : D^n \to \partial D^n$.*

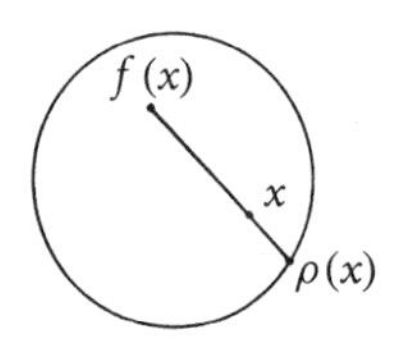

Figure 9.8.

The theorem and its corollary can actually be generalized to continuous maps through an additional argument (approximation of continuous maps by differentiable maps), and then the corollary is called the ***Brouwer fixed-point theorem***.

FOR EXERCISE 9.3. Up to now, we have used partitions of unity only to "partition" a differential form. But they are used far more often when individual pieces are given locally but don't fit together, to weld them into a smooth global object. The procedure is described in detail in Chapter VIII, §4 of [J:*Top*]. You don't have to study the details—skimming that section will give you the idea of the procedure in our Exercise 9.3.

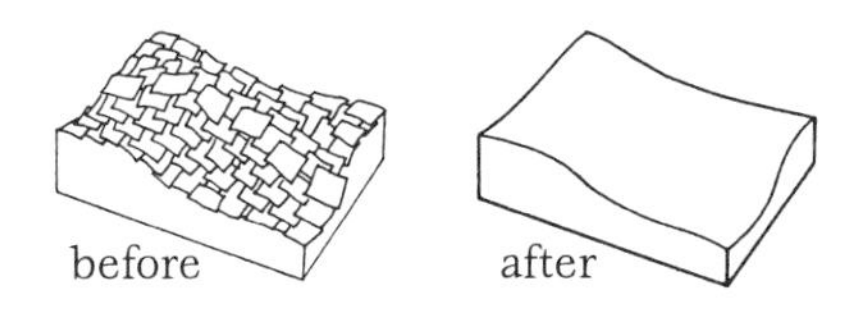

Figure 9.9.

For Exercise 9.4. Partitions of unity are an extremely convenient tool for constructing forms and functions. Here, too, you just have to solve the problem locally and then refer to partitions of unity in one or two cleverly worded lines.

CHAPTER 10 Classical Vector Analysis

10.1 Introduction

In hindsight, it is easy to say that classical nineteenth-century vector analysis is about the Cartan derivative and Stokes's theorem, though in a notation in which these objects are unrecognizable at first sight.

If we go from analysis on manifolds toward classical analysis, we can see from a distance that there we'll be dealing only with submanifolds of $\mathbb{R}^3$ or, at worst, $\mathbb{R}^n$. So? Our ideas can even be applied to arbitrary manifolds.

As we get closer, we also see that the integrands are usually defined not just on M but on a whole open neighborhood X of M, on $X = \mathbb{R}^3$ or $X = \mathbb{R}^3 \setminus \{0\}$, for instance, or something similar. So what? Surely our analysis on M can take care of any $\eta \in \Omega^k X$; we just apply $\iota : M \hookrightarrow X$ and consider the restriction $\iota^* \eta \in \Omega^k M$.

This is true in principle, but forms on open subsets X of $\mathbb{R}^3$ should *not* immediately be dismissed. For one thing, if we now enter classical vector analysis, we have to recognize the forms η on X as the real objects of interest. They describe physical "fields" of various kinds, while the submanifolds

$M \subset X$ are only called in as auxiliaries, to "test" an $\eta \in \Omega^k X$ in some sense—to analyze it. Think of a flux density $\eta \in \Omega^2 X$ on a region X in space, for instance, whose net flux $\int_M \eta$ through various surfaces M is to be considered.

But there are also technical advantages to computing with forms η on X once they're already given there, even if we could actually get by with the partial information $\iota^*\eta$. On X we have the canonical coordinates x^1, x^2, x^3 of $\mathbb{R}^3$, so we can use the dx^μ to represent the differential forms globally. Since the wedge product and the Cartan derivative are compatible with maps, in particular with inclusion ($\iota^*\omega \wedge \iota^*\eta = \iota^*(\omega \wedge \eta)$ and $d\iota^*\eta = \iota^* d\eta$), it makes no difference whether we compute before or after applying ι^*, and doing the computation first is often easier.

The reason classical vector analysis is completely unrecognizable at first sight as a domain of application of the Cartan calculus is the complete absence of differential forms. The concept isn't mentioned at all! The theory deals instead with vector fields on X—hence the name—and with the gradient, curl, and divergence operators. Only the fact that the integration is over volumes, surfaces, and curves indicates that there is, after all, a connection with analysis on manifolds.

This connection is made through the bases for the fields and forms in terms of the coordinates x^1, x^2, x^3. To be precise, 1-forms and 2-forms are described with respect to the bases by three component functions, *as are vector fields*, and 3-forms by one. The next section contains the details of this translation of forms into the language of classical vector analysis.

10.2 The Translation Isomorphisms

For an open subset X of $\mathbb{R}^3$, let $\mathcal{V}(X)$ denote the vector space of differentiable vector fields and $C^\infty(X)$ that of differentiable functions on X.

We denote the component functions of a vector field $\vec{a} \in \mathcal{V}(X)$ by a_1, a_2, a_3, with *sub*scripts, in deliberate contrast to the Ricci calculus. Otherwise a clash with the Ricci calculus would just occur somewhere else! This is an indication that describing 1-forms and 2-forms by vector fields is *not* in fact compatible with all changes of coordinates. But as long as we continue to use the canonical coordinates of $\mathbb{R}^3$, we may think of a vector field $\vec{a}$ as simply a triple $\vec{a} = (a_1, a_2, a_3)$ of functions.

In order to give transparent formulas for the translation isomorphisms, we introduce the following notation.

Definition. Let $X \subset \mathbb{R}^3$ be open. The $\mathbb{R}^3$-valued ("vector-valued") 1-form and 2-form

$$d\vec{s} := \begin{pmatrix} dx^1 \\ dx^2 \\ dx^3 \end{pmatrix} \in \Omega^1(X, \mathbb{R}^3) \text{ and } d\vec{S} := \begin{pmatrix} dx^2 \wedge dx^3 \\ dx^3 \wedge dx^1 \\ dx^1 \wedge dx^2 \end{pmatrix} \in \Omega^2(X, \mathbb{R}^3)$$

are called the ***vectorial line element*** and the ***vectorial area element***, respectively, and the ordinary real-valued 3-form

$$dV := dx^1 \wedge dx^2 \wedge dx^3 \in \Omega^3 X$$

is called the ***volume element*** of X.

Convention. The usual translation isomorphisms are given by

$$\begin{aligned} \mathcal{V}(X) &\xrightarrow{\cong} \Omega^1 X, & \vec{a} &\mapsto \vec{a} \cdot d\vec{s}, \\ \mathcal{V}(X) &\xrightarrow{\cong} \Omega^2 X, & \vec{b} &\mapsto \vec{b} \cdot d\vec{S}, \\ C^\infty(X) &\xrightarrow{\cong} \Omega^3 X, & c &\mapsto c\, dV. \end{aligned}$$

Here the dot denotes the standard scalar product on $\mathbb{R}^3$. But if we write $\vec{a}$, $\vec{b}$ as rows and $d\vec{s}$, $d\vec{S}$ as columns, the dot can also be read as the symbol for the matrix product.

The dictionary for translating classical vector analysis into the Cartan calculus, and vice versa, starts with this convention. As you can see, the translation from right to left is unambiguous, but whether a vector field has to be interpreted as a 1-form or a 2-form can't be inferred from the dictionary alone. Just as in foreign languages, it depends on the context.

Incidentally, the terms *line element, area element, and volume element* become plausible once we understand the geometric meaning of these forms.

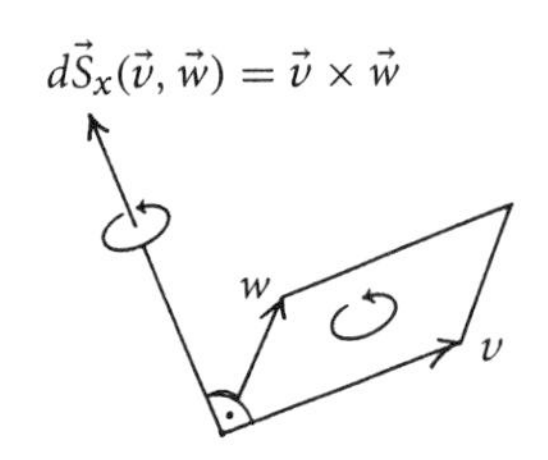

Figure 10.1. The cross product: a reminder

$$\vec{v} \times \vec{w} := \begin{pmatrix} v^2w^3 - v^3w^2 \\ v^3w^1 - v^1w^3 \\ v^1w^2 - v^2w^1 \end{pmatrix}$$

Note. *At every point* $x \in X$,

$$\begin{array}{rcll} d\vec{s}_x : \mathbb{R}^3 & \longrightarrow & \mathbb{R}^3 & \textit{is the identity,} \\ d\vec{S}_x : \mathbb{R}^3 \times \mathbb{R}^3 & \longrightarrow & \mathbb{R}^3 & \textit{is the cross product,} \\ dV_x : \mathbb{R}^3 \times \mathbb{R}^3 \times \mathbb{R}^3 & \longrightarrow & \mathbb{R}^3 & \textit{is the determinant.} \end{array}$$

By linearity, these assertions need be proved only for the canonical basis vectors, for which they are obvious (consider $d\vec{S}_x(\vec{e}_1, \vec{e}_2) = \vec{e}_3$ and look at cyclic permutations, as for the cross product). So let's go straight to the interpretation. The determinant gives the elementary-geometric volume of a positively oriented 3-span. The response of the cross product to an oriented 2-span is the normal vector whose length is the elementary-geometric area of the span (there are two such vectors) and whose direction is the one that, followed by the orientation of the span, gives the spatial orientation. The identity needs no explanation. If you imagine that forms respond to small ("infinitesimal") cells, you can see that the names make sense.

10.3 Gradient, Curl, and Divergence

Now let's use this dictionary to translate the Cartan derivative into the language of vector analysis. As always, $X \subset \mathbb{R}^3$ denotes an open subset. For $f \in C^\infty(X)$, we have

$$\begin{aligned} df &= \frac{\partial f}{\partial x^1}dx^1 + \frac{\partial f}{\partial x^2}dx^2 + \frac{\partial f}{\partial x^3}dx^3 \\ &= \left(\frac{\partial f}{\partial x^1}, \frac{\partial f}{\partial x^2}, \frac{\partial f}{\partial x^3}\right) \cdot d\vec{s}, \end{aligned}$$

and for vector fields $\vec{a}, \vec{b} \in \mathcal{V}(X)$ the Cartan derivatives of the 1-form $\vec{a} \cdot d\vec{s}$ and the 2-form $\vec{b} \cdot d\vec{S}$ are

$$\begin{aligned} d(\vec{a} \cdot d\vec{s}) &= d\sum_{\mu} a_\mu dx^\mu = \sum_{\mu,\nu} \partial_\nu a_\mu dx^\nu \wedge dx^\mu \\ &= (\partial_2 a_3 - \partial_3 a_2) dx^2 \wedge dx^3 + \text{ cyclic permutations} \\ &= (\partial_2 a_3 - \partial_3 a_2,\ \partial_3 a_1 - \partial_1 a_3,\ \partial_1 a_2 - \partial_2 a_1) \cdot d\vec{S} \end{aligned}$$

and

$$\begin{aligned} d(\vec{b} \cdot d\vec{S}) &= db_1 \wedge dx^2 \wedge dx^3 + \text{ cyclic permutations} \\ &= \frac{\partial b_1}{\partial x^1} dx^1 \wedge dx^2 \wedge dx^3 + \text{ cyclic permutations} \\ &= \left(\frac{\partial b_1}{\partial x^1} + \frac{\partial b_2}{\partial x^2} + \frac{\partial b_3}{\partial x^3} \right) dV. \end{aligned}$$

This is where we encounter the three classical operators of vector analysis, for which we fix our notation.

Definition. For $X \subset \mathbb{R}^3$ open, we define the ***gradient***, ***curl***, and ***divergence***

$$\begin{aligned} \operatorname{grad} : C^\infty(X) &\longrightarrow \mathcal{V}(X), \\ \operatorname{curl} : \mathcal{V}(X) &\longrightarrow \mathcal{V}(X), \\ \operatorname{div} : \mathcal{V}(X) &\longrightarrow C^\infty(X), \end{aligned}$$

by

$$\begin{aligned} \operatorname{grad} f &:= \left(\frac{\partial f}{\partial x^1}, \frac{\partial f}{\partial x^2}, \frac{\partial f}{\partial x^3} \right), \\ \operatorname{curl} \vec{a} &:= \left(\frac{\partial a_3}{\partial x^2} - \frac{\partial a_2}{\partial x^3}, \frac{\partial a_1}{\partial x^3} - \frac{\partial a_3}{\partial x^1}, \frac{\partial a_2}{\partial x^1} - \frac{\partial a_1}{\partial x^2} \right), \\ \operatorname{div} \vec{b} &:= \frac{\partial b_1}{\partial x^1} + \frac{\partial b_2}{\partial x^2} + \frac{\partial b_3}{\partial x^3}. \end{aligned}$$

The computations above for the translation of the Cartan derivative have given us the following result.

Note. *For* $X \subset \mathbb{R}^3$ *open,* $df = \operatorname{grad} f \cdot d\vec{s}$, $d(\vec{a} \cdot d\vec{s}) = \operatorname{curl} \vec{a} \cdot d\vec{S}$, *and* $d(\vec{b} \cdot d\vec{S}) = (\operatorname{div} \vec{b}) dV$. *Hence the diagram*

$$\begin{array}{ccccccccccc}
0 & \longrightarrow & \Omega^0 X & \xrightarrow{d} & \Omega^1 X & \xrightarrow{d} & \Omega^2 X & \xrightarrow{d} & \Omega^3 X & \longrightarrow & 0 \\
 & & = \uparrow & & \cong \uparrow & & \cong \uparrow & & \cong \uparrow & & \\
0 & \longrightarrow & C^\infty(X) & \underset{\operatorname{grad}}{\longrightarrow} & \mathcal{V}(X) & \underset{\operatorname{curl}}{\longrightarrow} & \mathcal{V}(X) & \underset{\operatorname{div}}{\longrightarrow} & C^\infty(X) & \longrightarrow & 0
\end{array}$$

is commutative.

Corollary. $\operatorname{curl} \operatorname{grad} f = 0$ *and* $\operatorname{div} \operatorname{curl} \vec{a} = 0$ *for all functions* f *and all vector fields* $\vec{a}$.

We pause at this stage of the translation to note how Stokes's theorem looks as a theorem about vector fields or functions on X. The corollary of Stokes's theorem that results for $\dim M = 3$ is called Gauss's integral theorem or the divergence theorem.

Gauss's Integral Theorem. *Let* $X \subset \mathbb{R}^3$ *be open and let* $\vec{b}$ *be a differentiable vector field on* X. *Then*

$$\int_{M^3} \operatorname{div} \vec{b} \, dV = \int_{\partial M^3} \vec{b} \cdot d\vec{S}$$

for all compact three-dimensional submanifolds-with-boundary $M^3 \subset X$.

Observe that three-dimensional submanifolds are canonically oriented by $\mathbb{R}^3$. In the two-dimensional case we have the classical Stokes's theorem, for which the more general theorem is named.

Stokes's Integral Theorem. *Let* $X \subset \mathbb{R}^3$ *be open and let* $\vec{a}$ *be a differentiable vector field on* X. *Then*

$$\int_{M^2} \operatorname{curl} \vec{a} \cdot d\vec{S} = \int_{\partial M^2} \vec{a} \cdot d\vec{s}$$

for all oriented compact two-dimensional submanifolds-with-boundary ("surfaces") $M^2 \subset X$.

For completeness, we also mention the one-dimensional case, although it has no name of its own.

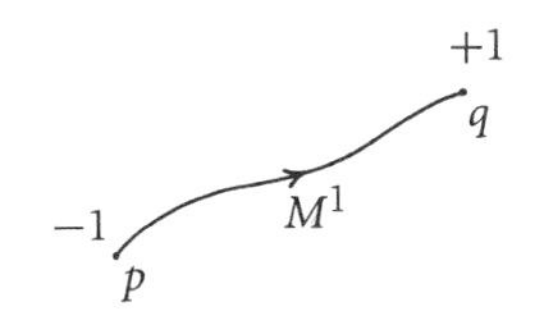

Figure 10.2.

If $X \subset \mathbb{R}^3$ is open and $f : X \to \mathbb{R}$ is a differentiable function, then

$$\int_{M^1} \operatorname{grad} f \cdot d\vec{s} = f(q) - f(p)$$

for all oriented compact one-dimensional submanifolds $M^1 \subset X$ from p to q.

10.4 Line and Area Elements

In the integral notation of classical vector analysis, the *nonvectorial* line element ds and the *nonvectorial* area element dS play a central role. So we introduce these two "elements" next, and to make them easier to understand we start with an interpretation too closely tied to differential calculus to be really authentic.

Definition. If $M \subset \mathbb{R}^n$ is an oriented k-dimensional submanifold, the k-form $\omega_M \in \Omega^k M$ that responds with $+1$ to every positively oriented orthonormal basis of a tangent space T_pM is called the ***canonical volume form*** of M. We call the canonical volume form the ***line element*** when $k = 1$ and the ***area element*** when $k = 2$, and denote it by ds and dS, respectively.

The intuitive meaning of the canonical volume form is clear, and we have already encountered it in the exercises (see Exercises 3.2 and 8.4). Its response to a positively oriented k-span is the elementary-geometric k-dimensional volume of the span. If we denote the k-dimensional volume of a set $A \subset M$ by $\mathrm{Vol}_k(A)$, then

$$\mathrm{Vol}_k(A) = \int_A \omega_M$$

whenever the integral exists—think of this equation as a definition if you have no other definition of the k-dimensional volume in $\mathbb{R}^n$ available, and as a lemma otherwise. In particular, $\int_A ds$ is the arc length of A for $k = 1$, and $\int_A dS$ is the area of A for $k = 2$. When $k = 3$ we can also write dV for the canonical volume element, and for $M^3 \subset \mathbb{R}^3$ this agrees with our earlier definition $dV = dx^1 \wedge dx^2 \wedge dx^3$.

But how are ds and dS related to the *vectorial* line and area elements $d\vec{s} \in \Omega^1(X, \mathbb{R}^3)$ and $d\vec{S} \in \Omega^2(X, \mathbb{R}^3)$ that appear in our dictionary (Section 10.2) and in the integral theorems? It is clear from the geometric meaning of $d\vec{s}$ and $d\vec{S}$ (see the note in Section 10.2) that in the two-dimensional case the responses of $\iota^* d\vec{S}$ and dS to an oriented tangential 2-span have the same *magnitude*, and similarly for $\iota^* d\vec{s}$ and ds in the one-dimensional case. But while ds and dS respond with real numbers, $d\vec{s}$ and $d\vec{S}$ respond with *vectors*; in fact, $d\vec{s}$ acts as the identity and gives a tangent vector, and $d\vec{S}$ acts as the cross product and gives a normal vector. In order to express this precisely and with the right sign, we introduce the following notation.

Notation. Let $M \subset \mathbb{R}^n$ be an oriented k-dimensional submanifold, $k = 1$ or $k = n - 1$.

(a) If $k = 1$, let $\vec{T} : M \to \mathbb{R}^n$ denote the positively oriented unit tangent field.

(b) If $k = n - 1$, let $\vec{N} : M \to \mathbb{R}^n$ denote the orienting unit normal field; that is, $\vec{N}(x) \perp T_xM$, $\|\vec{N}(x)\| = 1$, and $\vec{N}(x)$ followed by a positively oriented basis of T_xM gives a positively oriented basis of $\mathbb{R}^n$.

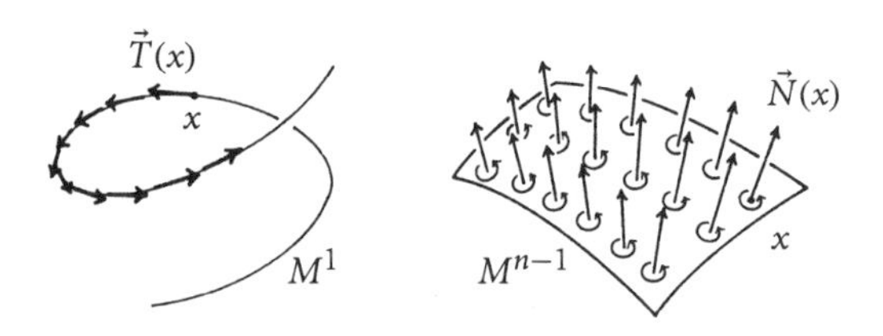

Figure 10.3. Unit tangent vector and unit normal vector in the dimension-1 and codimension-1 cases, respectively

Lemma. *Let $X \subset \mathbb{R}^3$ be open and let $\iota : M^k \hookrightarrow \mathbb{R}^3$, for $k = 1$, 2, be the inclusion of an oriented k-dimensional submanifold. Then*

$$\iota^* d\vec{s} = \vec{T} ds \in \Omega^1(M^1, \mathbb{R}^3) \quad \text{and} \quad \iota^* d\vec{S} = \vec{N} dS \in \Omega^2(M^2, \mathbb{R}^3).$$

PROOF. If $k = 1$, then $d\vec{s}(\vec{T}) = \vec{T}$ and $ds(\vec{T}) = 1$ at every point, so the first equation holds. If $k = 2$ and a positively oriented orthonormal basis $(\vec{v}, \vec{w})$ of $T_x M^2$ is given, then $\vec{N}(x)$ extends this basis to a positively oriented orthonormal basis $(\vec{N}, \vec{v}, \vec{w})$ of $\mathbb{R}^3$. Moreover, $dS(\vec{v}, \vec{w}) = 1$, so $\vec{N} dS(\vec{v}, \vec{w}) = \vec{N} = \vec{v} \times \vec{w} = d\vec{S}(\vec{v}, \vec{w})$. □

10.5 The Classical Integral Theorems

The nonvectorial line and area elements place the classical notation of the integral theorems at our disposal. We can now write the integral of a 1-form $\vec{a} \cdot d\vec{s}$ over an oriented one-dimensional submanifold as

$$\int_{M^1} \vec{a} \cdot d\vec{s} = \int_{M^1} \vec{a} \cdot \vec{T}\, ds,$$

where of course $\vec{a} \cdot \vec{T} : M^1 \to \mathbb{R}$ means the function on M^1 given by $x \mapsto \vec{a}(x) \cdot \vec{T}(x)$ (so $\vec{a} \cdot \vec{T}$ really means $(\vec{a}|M^1) \cdot \vec{T}$). Intuitively, this notation describes what happens to the vector field under integration, since $\vec{a}(x) \cdot \vec{T}(x) =: a_{\tan}(x)$ is the tangential component of the vector $\vec{a}(x)$ at the point $x \in M^1$, and the contribution to the integral of a little piece of M^1 near x is thus approximately the product $a_{\tan}(x)\Delta s$ of this tangential component and the arc length Δs of the little piece.

The two-dimensional case is similar:

$$\int_{M^2} \vec{b} \cdot d\vec{S} = \int_{M^2} \vec{b} \cdot \vec{N}\, dS,$$

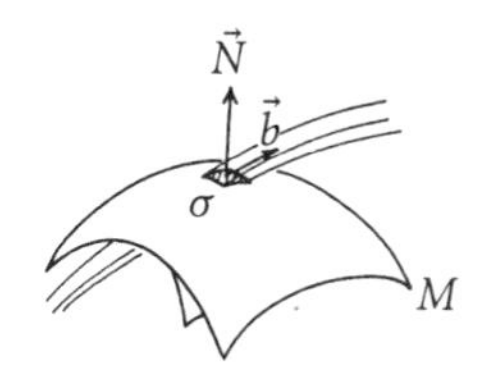

Figure 10.4. Part of the flux through the cell σ

where $\vec{b}(x) \cdot \vec{N}(x) =: b_{\text{nor}}(x)$ is now the normal component of $\vec{b}$ at the point x of the surface M^2. If $\vec{b}$ gives the strength and direction of a flux, for example, then $b_{\text{nor}} dS$ responds to a cell in M^2 with the rate of flow across M^2. In particular, we now obtain the two integral theorems of Gauss and Stokes (see Section 10.3) in what may be their most common version.

Gauss's Integral Theorem. *If $X \subset \mathbb{R}^3$ is open and $\vec{b}$ is a differentiable vector field on X, then*

$$\int_{M^3} \operatorname{div} \vec{b}\, dV = \int_{\partial M^3} \vec{b} \cdot \vec{N}\, dS$$

for all compact three-dimensional submanifolds-with-boundary $M^3 \subset X$.

Here M^3 is thought of as canonically oriented by $\mathbb{R}^3$, so by the orientation convention $\vec{N}$ means the *outward* unit normal vector field on ∂M.

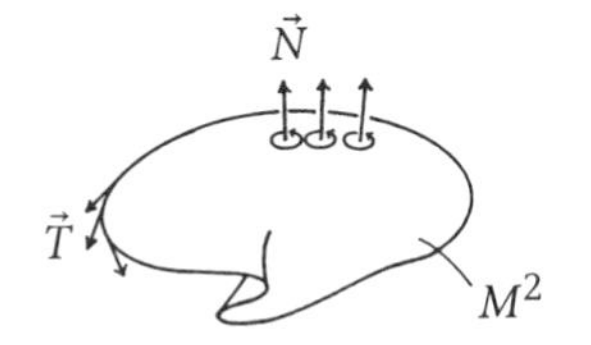

Figure 10.5. $\vec{T}$ and $\vec{N}$ for Stokes's theorem

Stokes's Integral Theorem. *If $X \subset \mathbb{R}^3$ is open and $\vec{a}$ is a differentiable vector field on X, then*

$$\int_{M^2} \operatorname{curl} \vec{a} \cdot \vec{N}\, dS = \int_{\partial M^2} \vec{a} \cdot \vec{T}\, ds$$

for all oriented compact surfaces-with-boundary $M^2 \subset X$.

As an application of Gauss's integral theorem we consider the special case $\vec{b} = \operatorname{grad} f$, where we have to take the volume integral of div grad f. Written in the coordinates x, y, z of $\mathbb{R}^3$, div grad is the familiar ***Laplace operator***, or ***Laplacian***, Δ:

$$\Delta f := \frac{\partial^2 f}{\partial x^2} + \frac{\partial^2 f}{\partial y^2} + \frac{\partial^2 f}{\partial z^2}.$$

In this context the notation ∇ ("***nabla***") tends to be used for the gradient:

$$\nabla f := \left(\frac{\partial f}{\partial x}, \frac{\partial f}{\partial y}, \frac{\partial f}{\partial z} \right).$$

Throughout what follows, let f and g be differentiable functions on an open set $X \subset \mathbb{R}^3$ and let $M^3 \subset X$ be a compact three-dimensional submanifold-with-boundary, as in Gauss's integral theorem. Setting $\vec{b} = \nabla f$ gives an immediate corollary.

Corollary 1.

$$\int_{M^3} \Delta f \, dV = \int_{\partial M^3} \nabla f \cdot \vec{N} dS.$$

Since $\nabla f \cdot \vec{N}$ is the directional derivative of f in the direction of the outward normal ($\vec{N} f$ if vectors are viewed as derivations), it is also written

$$\nabla f \cdot \vec{N} =: \frac{\partial f}{\partial n}$$

(the "normal derivative of f"), and Gauss's theorem for grad f takes the following form:

Corollary 2.

$$\int_{M^3} \Delta f \, dV = \int_{\partial M^3} \frac{\partial f}{\partial n} dS.$$

A bit more generally, we now set $\vec{b} = g \nabla f$. The ordinary product rule gives

$$\operatorname{div}(g \nabla f) = \nabla g \cdot \nabla f + g \Delta f,$$

and hence the following result.

Corollary 3 (Green's first identity).

$$\int_{M^3} (\nabla g \cdot \nabla f + g \Delta f) \, dV = \int_{\partial M^3} g \nabla f \cdot \vec{N} \, dS.$$

Since the scalar product $\nabla g \cdot \nabla f$ is symmetric in f and g, this gives another identity.

Corollary 4 (Green's second identity).

$$\begin{aligned}\int\limits_{M^3} (f\,\Delta g - g\Delta f)\,dV &= \int\limits_{\partial M^3} (f\nabla g - g\nabla f)\cdot\vec{N}\,dS\\ &= \int\limits_{\partial M^3} \left(f\frac{\partial g}{\partial n} - g\frac{\partial f}{\partial n}\right)\,dS.\end{aligned}$$

10.6 The Mean-Value Property of Harmonic Functions

Of course, such an enumeration of special cases is a rather dry affair unless something more is done with them. In physics these formulas come alive! We can't go into this, but we will derive another nice *mathematical* result from Gauss's theorem (or rather from its Corollary 1 in Section 10.5).

Definition. Let $X \subset \mathbb{R}^3$ be open. A differentiable function $f : X \to \mathbb{R}$ is called ***harmonic*** if $\Delta f \equiv 0$.

Theorem (Mean value property of harmonic functions). *Let $f : X \to \mathbb{R}$ be harmonic and let K be a closed ball lying entirely in X, with radius r, center p, and boundary S. Then*

$$f(p) = \frac{\int_S f\,dS}{\int_S dS} = \frac{1}{4\pi r^2}\int\limits_S f\,dS;$$

that is, the value of the function at the center is the mean value of the function on the surface of the ball.

Proof. Without loss of generality, let $p = 0$. When (x^1, x^2, x^3) occurs as a tangent vector, we write $\vec{x} := (x^1, x^2, x^3)$ for consistency with the vector-analytic formulas. We should actually distinguish between $x = (x^1, x^2, x^3)$ as a point in $M = \mathbb{R}^3$ and $\vec{x} = (x^1, x^2, x^3)$ as a vector in $T_q\mathbb{R}^3 \cong \mathbb{R}^3$, but we haven't inserted the distinction between $\mathbb{R}^n$ and its tangent spaces into our notation anywhere else.

For every $t \in [0,1]$, the function f_t defined by $f_t(\vec{x}) := f(t\vec{x})$ is also harmonic on some domain containing K and has the same value at p as f. Since the constant function f_0 obviously has the property $4\pi r^2 f(p) = \int_S f_0 dS$, it suffices to show that the integral

$$I_t := \int_S f_t dS$$

is independent of t, i.e. that $\frac{d}{dt} I_t \equiv 0$. Since $\frac{d}{dt} f(t\vec{x}) = \nabla f(t\vec{x}) \cdot \vec{x}$ and $\nabla f_t(\vec{x}) = t \nabla f(t\vec{x})$, we have

$$\frac{d}{dt} I_t = \frac{1}{t} \int_S \nabla f_t \cdot \vec{x}\, dS \quad \text{for } t > 0.$$

But on the boundary S of the ball of radius r, the outward normal is $\vec{N} = \vec{x}/r$; hence, by Corollary 1 of Section 10.5,

$$\frac{d}{dt} I_t = \frac{r}{t} \int_S \nabla f_t \cdot \vec{N}\, dS = \frac{r}{t} \int_K \Delta f_t \, dV,$$

and is therefore zero because f_t is a harmonic function. □

Corollary (Maximum principle for harmonic functions). *If $X \subset \mathbb{R}^3$ is open and connected, and if the harmonic function $f : X \to \mathbb{R}$ has an extremum, then f is constant.*

PROOF. Without loss of generality, let $f(x) \leq f(x_0) =: y_0$ for all $x \in X$. The set $f^{-1}(y_0)$ is closed in X because f is continuous. But it is also open by the mean value property: Let $p \in f^{-1}(y_0)$. Then f must be constant and equal to y_0 on the boundary S of *any* ball about p that lies completely in X; otherwise continuity would give $\int_S f\, dS < f(p) \int_S dS$. Thus the nonempty set $f^{-1}(y_0)$ is open and closed in the connected subspace $X \subset \mathbb{R}^3$, so $f^{-1}(y_0) = X$. □

Corollary (Uniqueness for the Dirichlet boundary value problem). *Let $M \subset \mathbb{R}^3$ be a compact three-dimensional submanifold-with-boundary and let $f, g : M \to \mathbb{R}$ be continuous* *functions that are harmonic on $M \setminus \partial M$. If f and g agree on the boundary, i.e. if $f|\partial M = g|\partial M$, then $f = g$ everywhere on M.*

PROOF. Without loss of generality, let M be nonempty and connected. As a continuous function on a compact set, $f - g$ must take on extrema. That is, there are $x_0, x_1 \in M$ with

$$f(x_0) - g(x_0) \leq f(x) - g(x) \leq f(x_1) - g(x_1)$$

for all $x \in M$. But either $f - g$ is already constant (in fact zero since $f|\partial M = g|\partial M$ and $\partial M \neq \emptyset$), or x_0 and x_1 must lie in ∂M, by the maximum principle applied to the harmonic function $f - g$. Thus, again using $f|\partial M = g|\partial M$, we find that $0 \leq f(x) - g(x) \leq 0$ for all $x \in M$. □

Our discussion of classical vector analysis has been restricted to the case $n = 3$. Some things are special to this dimension—the Cartan derivative (curl) changing 1-forms into $(n-1)$-forms, for instance—but others carry over to arbitrary n. In particular, on an open subset $X \subset \mathbb{R}^n$, 1-forms and n-forms are both translated into vector fields, and 0-forms and $(n-1)$-forms into functions; the Cartan derivatives $d : \Omega^0 X \to \Omega^1 X$ and $d : \Omega^{n-1} X \to \Omega^n X$ become the n-dimensional divergence and gradient, respectively; and the general Stokes's theorem gives the n-dimensional Gauss's theorem. This in turn implies the same Green's identities and the same theorems on harmonic functions as in the three-dimensional case.

10.7 The Area Element in the Coordinates of the Surface

After these examples of applications, we return to practical matters of vector analysis. How can we compute with line and area elements in local coordinates on the curve or the surface?

The integral of a k-form over the domain of an orientation-preserving chart on a k-dimensional manifold is simply the ordinary multiple integral of the downstairs component function, as we know from Chapter 5. How does this look concretely for the forms $\vec{a} \cdot d\vec{s}$ and $\vec{b} \cdot d\vec{S}$ of vector analysis? First, we should note as a peculiarity of the vector-analytic situation that the symbols x^1, x^2, x^3 are used for the coordinates of $\mathbb{R}^3$, so we have to choose others for the local coordinates of a surface $M^2 \subset \mathbb{R}^3$, say (u^1, u^2) or (u, v).

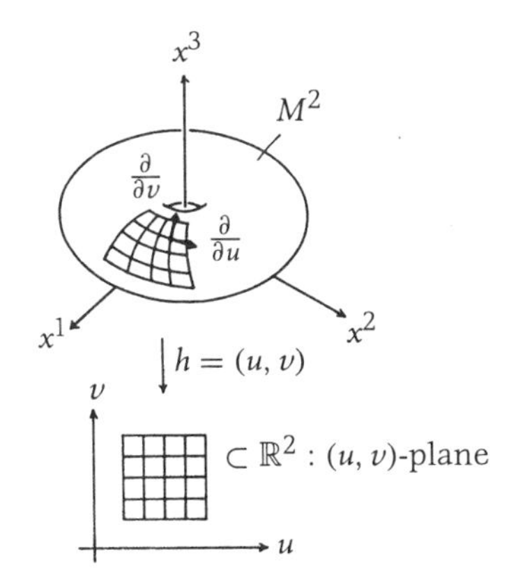

Figure 10.6. Notation for the coordinates

It is also more convenient in vector analysis to introduce coordinates on the surface "from below"; this means considering $\varphi := h^{-1}$ instead of h. Since $M^2 \subset \mathbb{R}^3$, φ is given by its three component functions $x^i = x^i(u, v)$, $i = 1, 2, 3$, abbreviated $\vec{x} = \vec{x}(u, v)$, on an open region (often denoted by $G \subset \mathbb{R}^2$) of the (u, v)-plane or half-plane. The canonical basis

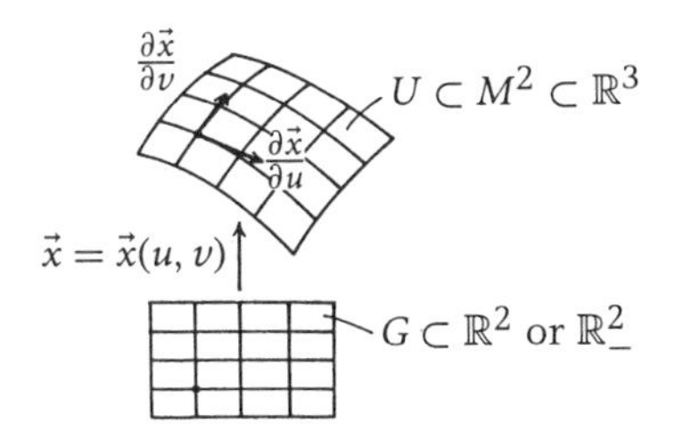

Figure 10.7. Local coordinates on a surface $M^2 \subset \mathbb{R}^3$

vectors of the chart, which we have written $\partial/\partial u$, $\partial/\partial v$ as elements of $T_pM \subset \mathbb{R}^3$, are then $\partial\vec{x}/\partial u$ and $\partial\vec{x}/\partial v$ as vectors in $\mathbb{R}^3$. Our next result comes from the definition and description of the area elements $d\vec{S}$ and dS (see Sections 10.2 and 10.4).

Corollary 1. *Let $M^2 \subset \mathbb{R}^3$ be an oriented surface in space and G an open region in $\mathbb{R}^2$ or $\mathbb{R}^2_-$. On G, let the inverse of an orientation-preserving chart (U, h) be given by $\vec{x} = \vec{x}(u, v)$. Then at every point $p = \vec{x}(u, v) \in U$ we have*

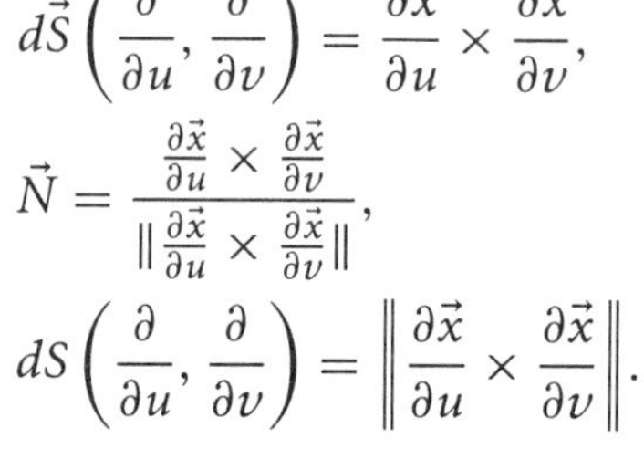

$$d\vec{S}\left(\frac{\partial}{\partial u}, \frac{\partial}{\partial v}\right) = \frac{\partial\vec{x}}{\partial u} \times \frac{\partial\vec{x}}{\partial v},$$

$$\vec{N} = \frac{\frac{\partial\vec{x}}{\partial u} \times \frac{\partial\vec{x}}{\partial v}}{\|\frac{\partial\vec{x}}{\partial u} \times \frac{\partial\vec{x}}{\partial v}\|},$$

$$dS\left(\frac{\partial}{\partial u}, \frac{\partial}{\partial v}\right) = \left\|\frac{\partial\vec{x}}{\partial u} \times \frac{\partial\vec{x}}{\partial v}\right\|.$$

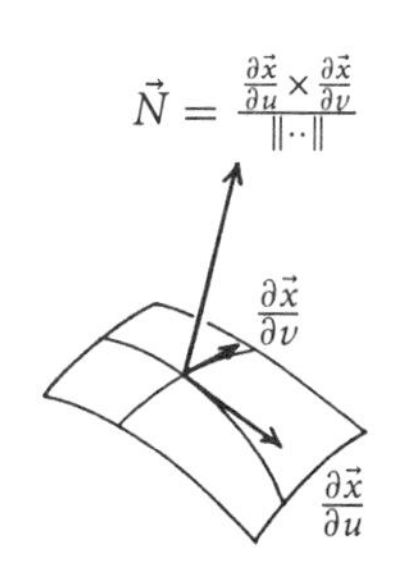

Figure 10.8. The orienting normal of the surface

Here again, the ambiguity of the notation u, v turns out to be useful. On the one hand, we can think of u, v as the coordinate functions on $U \subset M^2$, $\partial/\partial u$ and $\partial/\partial v$ as vector fields on U, and $du, dv \in \Omega^1 U$. With this interpretation,

$$d\vec{S}|U = \left(\frac{\partial\vec{x}}{\partial u} \times \frac{\partial\vec{x}}{\partial v}\right) du \wedge dv \in \Omega^2(U, \mathbb{R}^3)$$

or

$$dS|U = \left\|\frac{\partial\vec{x}}{\partial u} \times \frac{\partial\vec{x}}{\partial v}\right\| du \wedge dv \in \Omega^2 U$$

is the surface element as a 2-form on U. On the other hand, we can read u, v as the coordinates in G, and then

$$\left\|\frac{\partial\vec{x}}{\partial u} \times \frac{\partial\vec{x}}{\partial v}\right\|$$

on G is exactly the downstairs component function of dS. In particular, this gives the following corollary.

Corollary 2. *If, in addition to the hypotheses of Corollary 1, $\psi : M^2 \to \mathbb{R}$ is a function, then*

$$\int_U \psi \, dS = \iint_G \psi(\vec{x}(u,v)) \left\| \frac{\partial \vec{x}}{\partial u} \times \frac{\partial \vec{x}}{\partial v} \right\| du\, dv$$

whenever this double integral exists.

In vector analysis, the definition of the "surface integral" is usually based on this formula in coordinates, so it is important to observe that the same formula *also holds for an orientation-reversing chart* or a reorientation of U. Although a change of orientation reverses the sign of the integral when the integrand *stays the same*, our integrand doesn't stay the same at all—the area element dS changes sign under a change of orientation.

What we find for the line elements $d\vec{s}$ and ds is similar.

Note. *Let $M^1 \subset \mathbb{R}^3$ be an oriented curve (one-dimensional submanifold) in space, and on an interval $I \subset \mathbb{R}$ let the inverse of an orientation-preserving chart (U, h) be given by $t \mapsto \vec{x}(t)$. Then $d\vec{s}(\partial/\partial t) = \dot{\vec{x}}(t)$, $\vec{T} = \dot{\vec{x}}(t)/\|\dot{\vec{x}}(t)\|$, and $ds(\partial/\partial t) = \|\dot{\vec{x}}(t)\|$* at every point $p = \vec{x}(t) \in U$.

Hence

$$d\vec{s}\,|U = \dot{\vec{x}}\, dt \in \Omega^1(U, \mathbb{R}^3) \quad \text{or} \quad ds|U = \|\dot{\vec{x}}\|\, dt \in \Omega^1 U$$

is the line element as a 1-form on U. In particular,

$$ds = \sqrt{\dot{x}^1(t)^2 + \dot{x}^2(t)^2 + \dot{x}^3(t)^2}\, dt$$

at any point $\vec{x}(t)$, and we have an analogue of Corollary 2 for the line integral.

Corollary 3. *Since the line element ds changes sign under any change of orientation of U,*

$$\int_U \psi \, ds = \int_I \psi(\vec{x}(t)) \|\dot{\vec{x}}(t)\| \, dt$$

for any orientation of U.

Don't begrudge learning a third version of the classical Stokes's theorem, in addition to those in Sections 10.3 and 10.5. There's something special about this one. For this version, we let G denote a bounded region of the (u, v)-plane with smooth boundary, in our language a compact two-dimensional submanifold-with-boundary of $\mathbb{R}^2$. The boundary of G consists of one or several, say r, closed curves, which are oriented according to the orientation convention. Let them be parametrized consistently with the orientation convention by simple closed curves

$$\gamma_i : [\alpha_i, \beta_i] \to \partial G, \quad t \mapsto (u_i(t), v_i(t)), \quad i = 1, \ldots, r.$$

Corollary 4. *If $X \subset \mathbb{R}^3$ is open and $\vec{a}$ is a differentiable vector field on X, then for any differentiable map $\vec{x} = \vec{x}(u, v)$ from G to X we have*

$$\iint_G (\operatorname{curl} \vec{a}(\vec{x}(u, v))) \cdot \left(\frac{\partial \vec{x}}{\partial u} \times \frac{\partial \vec{x}}{\partial v} \right) du\, dv$$

$$= \sum_{i=1}^{r} \int_{\alpha_i}^{\beta_i} \vec{a}(\vec{x}(u_i(t), v_i(t)) \cdot \frac{d}{dt} \vec{x}(u_i(t), v_i(t)) dt.$$

The promised special feature of this version of the theorem is that the map $G \to X$, far from having to be an embedding (a diffeomorphism onto a submanifold $M \subset X$), may be *any* differentiable map $\varphi : G \to X$, even one that carries G into X crumpled up, singular, and self-intersecting! No new theorem in need of proof is hiding behind this, just the application of the general Stokes's theorem to G instead of to some $M \subset X$. More precisely, if we set $\omega := \vec{a} \cdot d\vec{s} \in \Omega^1 X$, then the formula of Corollary 4 just says that $\int_G d(\varphi^* \omega) = \int_{\partial G} \varphi^* \omega$.

10.8 The Area Element of the Graph of a Function of Two Variables

The special case where U is the graph of a differentiable function $z = z(x, y)$ is of particular interest. Then $u := x|U$ and $v := y|U$ are the coordinates of the canonical chart h. The inverse chart or "parametric representation" $G \xrightarrow{\varphi} U$ is then given by $x = x$, $y = y$, and $z = z(x, y)$, so the tangential basis vectors are

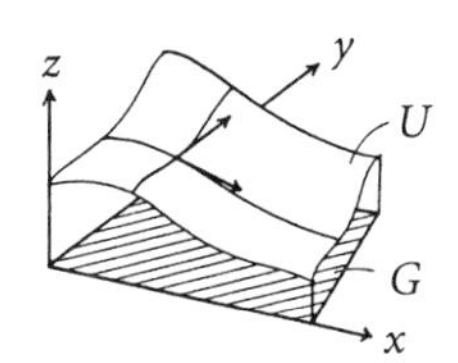

Figure 10.9. Basis for the tangent space to the graph

$$\frac{\partial \varphi}{\partial x} = \begin{pmatrix} 1 \\ 0 \\ \frac{\partial z}{\partial x} \end{pmatrix} \quad \text{and} \quad \frac{\partial \varphi}{\partial y} = \begin{pmatrix} 0 \\ 1 \\ \frac{\partial z}{\partial y} \end{pmatrix},$$

and the magnitude of their cross product is

$$\left\| \frac{\partial \varphi}{\partial x} \times \frac{\partial \varphi}{\partial y} \right\| = \sqrt{1 + \left(\frac{\partial z}{\partial x}\right)^2 + \left(\frac{\partial z}{\partial x}\right)^2}.$$

Corollary. *For a function $\psi : U \to \mathbb{R}$ on the graph*

$$U := \{(x, y, z(x, y)) : (x, y) \in G\}$$

of a differentiable function $z = z(x, y)$ on an open subset G of $\mathbb{R}^2$ or $\mathbb{R}^2_-$, the equation

$$\int\limits_U \psi \, dS = \iint\limits_G \psi(x, y, z(x, y)) \sqrt{1 + \left(\frac{\partial z}{\partial x}\right)^2 + \left(\frac{\partial z}{\partial y}\right)^2} \, dx \, dy$$

holds and is independent of orientation, whenever the double integral exists. In particular, the surface area of U is

$$\mathrm{Vol}_2(U) = \iint\limits_G \sqrt{1 + \left(\frac{\partial z}{\partial x}\right)^2 + \left(\frac{\partial z}{\partial y}\right)^2} \, dx \, dy.$$

10.9 The Concept of the Integral in Classical Vector Analysis

At the very end of this chapter we return yet again to the question of how, for its part, classical vector analysis really interprets integration on manifolds—essentially the surface integral—when it makes no use of differential forms. I conceded earlier that our interpretation of the line and area elements ds and dS as the canonical volume forms of oriented curves and surfaces is not quite authentic. What should the authentic interpretation be?

The genuine unadulterated area element dS of classical vector analysis is defined for every surface in space (analogously for every k-dimensional submanifold M^k of $\mathbb{R}^n$), and has absolutely nothing to do with orientation or orientability. But to each $p \in M$ it assigns, not an alternating 2-form, but a *density*

$$dS_p : T_pM \times T_pM \longrightarrow \mathbb{R}_+$$

(see Section 5.1); to be precise, dS_p just responds to a pair $(\vec{v}, \vec{w})$ of tangent vectors at the point p with the elementary-geometric area of the parallelogram they span, so

$$dS_p(\vec{v}, \vec{w}) = \|\vec{v} \times \vec{w}\|$$

for $M^2 \subset \mathbb{R}^3$. If M is actually oriented, then the volume form ω_M is related to the area element dS by

$$dS(\vec{v}, \vec{w}) = |\omega_M(v, w)|.$$

If we now imagine, as in Section 5.2, that dS responds in approximately this way to small cells, then for a function $f : M \to \mathbb{R}$ on an arbitrary surface (not necessarily oriented or even orientable), the intuitive meaning of

$$\int_M f \, dS$$

is clear. A formal definition is based on the local formula

$$\int_U f\,dS = \iint_G f\,(\vec{x}(u,v)) \cdot \left\| \frac{\partial \vec{x}}{\partial u} \times \frac{\partial \vec{x}}{\partial v} \right\| du\,dv.$$

(See Corollary 2 in Section 10.7.) If the Lebesgue integral for $\mathbb{R}^2$ is available, the general definition of integrability and the integral $\int_M f\,dS$ can be applied as in Section 5.4 by decomposing M into small measurable sets, with the additional convenience of not having to worry about whether the charts used are orientation-preserving. Incidentally,

$$A \longmapsto \int_A dS \in [0, \infty]$$

then gives a measure μ_M on the σ-algebra of measurable subsets of M (see Section 5.5), and $\int_M \ldots dS$ is just the Lebesgue integral on this measure space.

If one doesn't want to resort to the Lebesgue integral, partitions of unity (as in Section 9.6) offer an elegant way to base the definition of the surface integral on any notion, however modest, of the multiple integral (here, the double integral). If M is oriented, then $\int_M f\,dS$ according to this definition agrees with $\int_M f\,\omega_M$ according to the old definition.

This idea of the area element as given by the ordinary unoriented surface area is certainly the more obvious and the more elementary. It has the advantage of being directly applicable in the nonorientable case as well, without additional arguments. But it has the drawback of densities: the integrands $f\,dS$ are not differential forms and therefore cannot be inserted casually into the Cartan calculus. Orientation (crucial for the integral theorems) then appears in the form of the orienting normal field $\vec{N}$. This way of encoding the orientation hardly looks promising for generalization to arbitrary manifolds—even to, say, surfaces $M^2 \subset \mathbb{R}^4$. But whatever you may think of the differences between the old area element dS and the volume form $\omega_M \in \Omega^2 M$, I hope at least to have made these differences completely clear.

Yet even this description of the classical vector-analytic notion of the surface integral has been spruced up. The textbooks, still used to some extent today, that present classical vector analysis classically use neither the Lebesgue integral nor partitions of unity. The reader is led toward the definition of the surface integral in two stages: first, a plausibility argument showing that

$$\int_U f\, dS = \iint_G f(\vec{x}(u,v)) \left\| \frac{\partial \vec{x}}{\partial u} \times \frac{\partial \vec{x}}{\partial v} \right\| du\, dv$$

is the right formula locally, and second, instructions to cut the surface into appropriate "pieces," to each of which the formula can be applied. In view of the available notion of the double integral, this involves certain ad hoc conditions on the piecewise smoothness of the boundaries of these pieces of surface. Asking for clean definitions and proofs is not allowed. Even the question "What is a surface, anyway?" rarely gets a proper answer. The area element is given in the classical notation as

$$dS = \left\| \frac{\partial \vec{x}}{\partial u} \times \frac{\partial \vec{x}}{\partial v} \right\| du\, dv,$$

and all the reader learns about its status as a mathematical object is that it's an "expression," a "symbol." But this information, at best acceptable if a bit bald, is immediately superseded. This symbol is now converted to other coordinates, brought into a different form—an equality sign between "symbols" that look completely unalike? The putative proofs that the surface can be cut up and that the integral is well defined are just sketches of proofs, in fact sketches that, if actually carried out, would produce monstrosities.

Conceptually and technically, classical vector analysis is not only much narrower, but even within this narrower realm much clumsier, than analysis on manifolds. Anyone who uses it only for integration over the surface of a sphere or a cylinder and whose scientific interest is directed toward something completely different, namely the physical *content* of the equations, can of course be served quite well by a plausible, computable formula. But if, as a *mathematician*, you would like to understand the *structure* of vector analysis, then you should not expect much from those textbooks that, though rooted in the nineteenth century, are still going their stately way through this one.

10.10 Test

(1) To which vector field $\vec{v} = (v_1, v_2, v_3)$ in $\mathbb{R}^3$ does the 2-form $x dz \wedge dy$ correspond?

☐ $\vec{v} = (-x, 0, 0)$. ☐ $\vec{v} = (0, x, -x)$. ☐ $\vec{v} = (0, -x, x)$.

(2) As usual, let $r : \mathbb{R}^3 \setminus \{0\} \to \mathbb{R}$ denote distance from the origin. To which 1-form $\omega \in \Omega^1(\mathbb{R}^3 \setminus \{0\})$ does the outward radial unit vector field correspond?

☐ $\omega = dr$. ☐ $\omega = \dfrac{dr}{r}$. ☐ $\omega = \dfrac{dr}{r^2}$.

(3) It is always true that

☐ $\operatorname{grad}\operatorname{curl} = 0$. ☐ $\operatorname{curl}\operatorname{grad} = 0$. ☐ $\operatorname{div}\operatorname{grad} = 0$.

(4) In Stokes's integral theorem for a vector field $\vec{v}$, what should replace the dots in $\int_{M^2} \ldots dS = \int_{\partial M^2} \vec{v} \cdot \vec{T} ds$?

☐ $\operatorname{curl}\vec{v} \times \vec{N}$. ☐ $\operatorname{curl}\vec{v} \cdot \vec{N}$. ☐ $\|\operatorname{curl}\vec{v}\|$.

(5) The notation $\nabla \times \vec{v}$ of classical vector analysis, read sympathetically, can no doubt mean only

☐ $\operatorname{curl}\operatorname{div}\vec{v}$. ☐ $\operatorname{grad}\operatorname{div}\vec{v}$. ☐ $\operatorname{curl}\vec{v}$.

(6) According to Gauss's integral theorem, for every differentiable function $f : D^3 \to \mathbb{R}$ we have

☐ $\int_{D^3} \Delta f \, dV = \int_{S^2} f \, dS$.

□ $\int_{D^3} \Delta f \, dV = \int_{S^2} \frac{\partial f}{\partial r} \, dS$.

□ $\int_{D^3} \Delta f \, dV = \int_{S^2} \nabla f \, dS$.

(7) Let $f : X \to \mathbb{R}$ be constant on the boundary ∂M of the three-dimensional submanifold $M \subset X$. What does this mean for the normal derivative?

□ $\dfrac{\partial f}{\partial n} = 0$. □ $\dfrac{\partial f}{\partial n} = \nabla f$. □ $\dfrac{\partial f}{\partial n} = \pm\|\nabla f\|$.

(8) Let $X \subset \mathbb{R}^3$ be open and $f : X \to \mathbb{R}$ differentiable. Clearly, the formula $\int_{t_0}^{t_1} f'(\gamma(t))\dot{\gamma}(t)dt = f(q) - f(p)$ holds for any curve $\gamma : [t_0, t_1] \to X$ from p to q. But in what sense is this a special case of Stokes's theorem $\int_M d\omega = \int_{\partial M} \omega$? Set

□ $M := X$ and $\omega := f$.

□ $M := [t_0, t_1]$ and $\omega := \gamma^* f$.

□ $M := [t_0, t_1]$ and $\omega := f$.

(9) In terms of the coordinate x, what is the line element ds of the graph $\{(x, \sqrt{x}) : x > 0\}$?

□ $\sqrt{1 + \dfrac{1}{4x}}\, dx$. □ $\sqrt{1 + \dfrac{1}{2\sqrt{x}}}\, dx$. □ $\sqrt{x^2 + x}\, dx$.

(10) What is the area element $dS \in \Omega^2 S^2$ of the 2-sphere with the usual orientation, expressed in terms of the geographic angular coordinates (eastern) longitude λ and (northern) latitude β?

□ $\sin\beta \, d\beta \wedge d\lambda$. □ $\cos\beta \, d\lambda \wedge d\beta$. □ $\sin\beta \, d\lambda \wedge d\beta$.

10.11 Exercises

EXERCISE 10.1. Let $X \subset \mathbb{R}^3$ be open. Let $\mathcal{V}(X) \cong \Omega^1 X \cong \Omega^2 X$ and $\Omega^0 X \cong \Omega^3 X$ be the isomorphisms established by the basis vector fields and basis forms

$$\begin{aligned} &\partial_1,\ \partial_2,\ \partial_3 \quad \text{for} \quad \mathcal{V}(X), \\ &dx^1,\ dx^2,\ dx^3 \quad \text{for} \quad \Omega^1 X, \\ &dx^2 \wedge dx^3,\ dx^3 \wedge dx^1,\ dx^1 \wedge dx^2 \quad \text{for} \quad \Omega^2 X, \\ \text{and} \quad &dx^1 \wedge dx^2 \wedge dx^3 \quad \text{for} \quad \Omega^3 X. \end{aligned}$$

If 1-forms and 2-forms are described in this way by vector fields and 3-forms by functions, what happens to the exterior product?

EXERCISE 10.2. Let $X \subset \mathbb{R}^3$ be open. For differentiable functions f and vector fields $\vec{v}$ and $\vec{w}$ on X, find vector-analytic product formulas for

(a) $\operatorname{curl}(f\vec{v}) =?$

(b) $\operatorname{div}(f\vec{v}) =?$

(c) $\operatorname{div}(\vec{v} \times \vec{w}) =?$

by translating them into the calculus of differential forms.

EXERCISE 10.3. Let $M \subset \mathbb{R}^2$ be a compact two-dimensional submanifold-with-boundary.

(a) Prove Green's theorem:

$$\int_{\partial M} f\,dx + g\,dy = \int_M \left(\frac{\partial g}{\partial x} - \frac{\partial f}{\partial y}\right) dx\,dy.$$

(b) What is the geometric meaning of the integrals $\int_{\partial M} x\,dy$ and $\int_{\partial M} y\,dx$?

EXERCISE 10.4. Recall that $\int_{D^3} dV = \frac{1}{3}\int_{S^2} dS$. Find and prove a generalization of this formula for D^n and S^{n-1}, $n \geq 1$.

EXERCISE 10.5. Let $M \subset \mathbb{R}^3$ be a compact three-dimensional submanifold-with-boundary and let $\vec{p}_1, \ldots, \vec{p}_n \in M \backslash \partial M$. Find

$$\sum_{k=1}^{n} \int_{\partial M} \frac{\vec{x} - \vec{p}_k}{|\vec{x} - \vec{p}_k|^3} \cdot d\vec{S}.$$

10.12 Hints for the Exercises

FOR EXERCISE 10.1. In Section 10.3 we "translated" the Cartan derivative into classical vector analysis. You should do that here for the wedge product; that is, convert the three maps

$$\begin{aligned} \Omega^1 X \times \Omega^1 X &\longrightarrow \Omega^2 X, \\ \Omega^1 X \times \Omega^2 X &\longrightarrow \Omega^3 X, \\ \Omega^1 X \times \Omega^1 X \times \Omega^1 X &\longrightarrow \Omega^3 X \end{aligned}$$

defined by the wedge product into corresponding relations among vector fields. For the first line, for example, find the map that makes the following diagram commutative:

$$\begin{array}{ccc} \Omega^1 X \times \Omega^1 X & \xrightarrow{\wedge} & \Omega^2 X \\ \uparrow \cong & & \uparrow \cong \\ \mathcal{V}(X) \times \mathcal{V}(X) & \longrightarrow & \mathcal{V}(X) \end{array}$$

Of course, you can *calculate* this completely formally, but you should also do the best you can to *see* what the answer means.

FOR EXERCISE 10.2. This extends Exercise 10.1 and uses its results. In (a), for instance, we have to consider the diagram

$$\begin{array}{ccccc} \Omega^0 X \times \Omega^1 X & \xrightarrow{\wedge} & \Omega^1 X & \xrightarrow{d} & \Omega^2 X \\ \uparrow \cong & & \uparrow \cong & & \uparrow \cong \\ C^\infty(X) \times \mathcal{V}(X) & \xrightarrow{\cdot} & \mathcal{V}(X) & \xrightarrow{\text{curl}} & \mathcal{V}(X) \end{array}$$

where the vertical arrows represent the usual translation isomorphisms. On the top line we know what's going on: in the calculus of differential forms *one* product formula suffices, and for $\omega \in \Omega^r X$ it always reads $d(\omega \wedge \eta) = d\omega \wedge \eta + (-1)^r \omega \wedge d\eta$.

FOR EXERCISE 10.3. Here f and g are to be viewed as, say, differentiable in an open neighborhood X of M in $\mathbb{R}^2$. Of course, the exercise is in some way an application of Stokes's theorem, and we do have an integral of the form $\int_{\partial M} \omega$ on the left-hand side of (a). But notice that we *don't* have $\int_M d\omega$ on the right-hand side: $dx\,dy$ is *not* a typo for $dx \wedge dy$! One of the things this exercise requires is that you refer to the definition of the integral of a 2-form in this special situation. In both parts of the exercise you also have to pay attention to the sign!

FOR EXERCISE 10.4. What vector field $\vec{b}$ should you choose on a neighborhood of D^3 so that the formula in the exercise becomes exactly the statement $\int_{D^3} \operatorname{div} \vec{b} = \int_{\partial D^3} \vec{b} \cdot \vec{N} dS$ of Gauss's integral theorem? Once you've found this $\vec{b}$, even the generalization is completely obvious.

FOR EXERCISE 10.5. The physicists among you will recognize the integrand: $\vec{x}/r^3$ is the negative of the gradient of the harmonic function $1/r$. Anyone learning this here for the first time should do the calculation once. This makes the exercise an application of the formula

$$\int_{M^3} \Delta f \, dV = \int_{\partial M^3} \nabla f \cdot d\vec{S}$$

(Corollary 1 of Section 10.5). But not a direct application, because our integrand has isolated singularities! The best thing is to surround them with small balls, as is done for the residue theorem in complex function theory.

11 CHAPTER De Rham Cohomology

11.1 Definition of the de Rham Functor

We turn now from classical vector analysis to a completely different aspect of the calculus of differential forms. Consider the de Rham complex

$$0 \to \Omega^0 M \xrightarrow{d} \Omega^1 M \xrightarrow{d} \cdots$$

of a manifold M. The property $d \circ d = 0$ means that

$$\operatorname{im}(d : \Omega^{k-1} M \to \Omega^k M) \subset \ker(d : \Omega^k M \to \Omega^{k+1} M)$$

for every k, so we can take the quotient of these two vector spaces.

Definition. If M is a manifold, the quotient vector space

$$H^k M := \frac{\ker(d : \Omega^k M \to \Omega^{k+1} M)}{\operatorname{im}(d : \Omega^{k-1} M \to \Omega^k M)}$$

is called the kth ***de Rham cohomology group*** of M. The Cartan derivative d is also called the ***coboundary operator***; the differential forms in the image of some d are called

coboundaries, and those in the kernel ***cocycles***. If $\eta \in \Omega^k M$ is a k-dimensional cocycle, its coset

$$[\eta] := \eta + d\,\Omega^{k-1}M \in H^k M$$

is called the ***cohomology class*** of η.

The terms *boundaries, cycles,* and *homology class* come from *homology theory*, where a "chain" c has a "boundary" dc and is called a "cycle" if this boundary vanishes. Two cycles are called homologous if they differ only by a boundary. All this was already discussed in Sections 7.6 and 7.7.

In terms of geometric meaning, taking boundaries in homology theory corresponds to taking boundaries of compact manifolds-with-boundary. Since the Cartan derivative is dual to taking boundaries, in the sense that the effect of $d\alpha$ is just the effect of α on the boundary (as described in detail in Section 7.3), the terminology "coboundary operator" for d becomes understandable.

Lemma and Definition. *The wedge product and the functorial properties of the de Rham complex turn*

$$H^* := \bigoplus_{k=0}^{\infty} H^k$$

canonically into a contravariant functor from the differentiable category to the category of anticommutative graded algebras and their homomorphisms. This functor H^* is called simply the ***de Rham cohomology***.

Proof. We first show that the wedge product $H^r M \times H^s M \xrightarrow{\wedge} H^{r+s}M$ is well defined by

$$[\omega] \wedge [\eta] := [\omega \wedge \eta].$$

Clearly $\omega \wedge \eta$ is a cocycle whenever ω and η are: if $d\omega = 0$ and $d\eta = 0$, then $d(\omega\wedge\eta) = d\omega\wedge\eta + (-1)^r \omega\wedge d\eta = 0$. Without loss of generality, we need only check whether

$$[(\omega + d\alpha) \wedge \eta] = [\omega \wedge \eta],$$

so that $d\alpha \wedge \eta$ is a coboundary. But

$$d(\alpha \wedge \eta) = d\alpha \wedge \eta + (-1)^{r-1}\alpha \wedge d\eta = d\alpha \wedge \eta$$

because $d\eta = 0$. Hence the wedge product of a coboundary with a cocycle is always a coboundary, so the wedge product is also well defined for cohomology classes.

Furthermore, if $f : M \to N$ is a differentiable map, then the naturality of d (see Section 8.6) immediately implies that

$$\begin{aligned} f^* : H^k N &\longrightarrow H^k M, \\ [\eta] &\longmapsto [f^*\eta] \end{aligned}$$

is well defined. The algebraic and functorial properties now carry over from Ω^* (see Section 8.7) to H^*. □

11.2 A Few Properties

What can we say offhand about the computation of de Rham cohomology? First, a completely trivial observation:

Note. *If M is an n-dimensional manifold and $k > n$, then $\Omega^k M = 0$ and hence $H^k M = 0$.*

Of course, we also know the 0-cocycles, the functions $f \in \Omega^0 M$ with $df = 0$. These are the locally constant real functions, and the only coboundary among them is the zero function:

Note. *$H^0 M$ is the vector space of locally constant functions. In particular, if M is connected, then $H^0 M = \mathbb{R}$ canonically.*

In addition, Stokes's theorem gives us a statement about the other end of the de Rham sequence.

Corollary of Stokes's Theorem. *If M is an orientable closed (i.e. compact and without boundary) n-dimensional manifold, then $H^n M \neq 0$.*

PROOF. Orient M and choose $\eta \in \Omega^n M$ with $\int_M \eta \neq 0$ (using a chart and a bump function, for instance). Since $\Omega^{n+1} M = 0$, any n-form, and η in particular, is a cocycle. But

η is not a coboundary $d\omega$. If it were, the hypothesis $\partial M = \emptyset$ would imply $\int_M \eta = \int_M d\omega = \int_{\partial M} \omega = 0$ by Stokes's theorem. Hence $[\eta] \neq 0$ is in $H^n M$. □

Finally, if we look at the morphisms, then in addition to the functorial property of $f^* = H^k f : H^k N \to H^k M$ we can note the following:

Note. *If $f : M \to N$ is constant, then $H^k f = 0$ for all $k > 0$. If M and N are connected, then $H^0 f : \mathbb{R} \to \mathbb{R}$ is the identity for every f.*

So much for the meager results of direct inspection. In the next two sections, we establish an important nontrivial property: the homotopy invariance of de Rham cohomology.

Definition. Let M and N be differentiable manifolds. Two differentiable maps $f, g : M \to N$ are called ***differentiably homotopic*** if there is a ***differentiable homotopy*** h between them, that is, a differentiable map

$$h : [0, 1] \times M \longrightarrow N$$

such that $h(0, x) = f(x)$ and $h(1, x) = g(x)$ for all $x \in M$.

Since M and N (as always) may have boundary, we note explicitly that we will call a map $\varphi : U \to \mathbb{R}^n$ defined in an open subset U of $[0, 1] \times \mathbb{R}^n_-$ *differentiable* if every $u \in U$ has an open neighborhood V_u in $\mathbb{R}^{n+1}$ to which $\varphi | U \cap V_u$ can be extended differentiably.

Theorem (Homotopy invariance of de Rham cohomology). *If M, N are manifolds and $f, g : M \to N$ are differentiably homotopic maps, then*

$$f^* = g^* : H^k N \longrightarrow H^k M \quad \textit{for all } k.$$

Anything else I would really like to say now about the concept of homotopy in general, and the significance of homotopy invariance for functors from the geometric to the algebraic category in particular, can be found in [J:*Top*], Chapter V.

11.3 Homotopy Invariance: Looking for the Idea of the Proof

I would like to show you not only how the proof looks but also how to *find* it. Let ω be a k-dimensional cocycle on N. We have to show that

$$[f^*\omega] = [g^*\omega] \in H^k M,$$

or, in words, that the two cocycles $f^*\omega$ and $g^*\omega$ differ only by a coboundary $d\alpha$. So what we're *looking for* is an $\alpha \in \Omega^{k-1}M$ with

$$g^*\omega - f^*\omega = d\alpha.$$

So much for the problem. Now we inspect our tools. The only hypothesis is the existence of a differentiable homotopy between f and g, i.e. a differentiable map h from the cylinder $[0,1] \times M$ over M to N that coincides with f on the bottom $\{0\} \times M$ and with g on the top $\{1\} \times M$. Stated a bit more formally, what we have is

$$h \circ \iota_0 = f,$$
$$h \circ \iota_1 = g,$$

where $\iota_t : M \hookrightarrow [0,1] \times M$ denotes the inclusion at height t defined by $\iota_t(x) := (t, x)$.

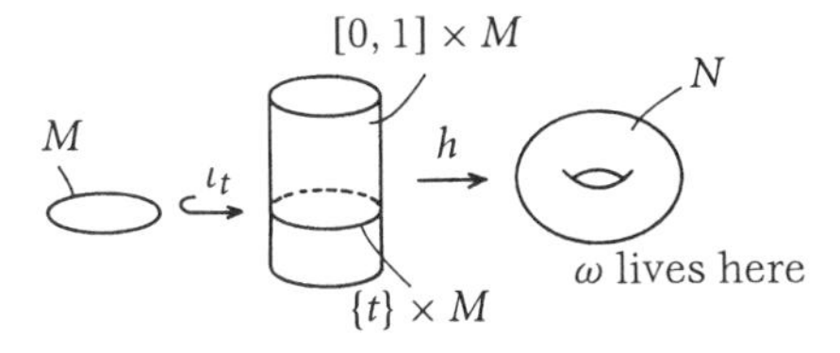

Figure 11.1. The homotopy h between $h \circ \iota_0 = f$ and $h \circ \iota_1 = g$

Then the induced cocycle $h^*\omega$ also coincides with $f^*\omega$ on the bottom and with $g^*\omega$ on the top, or more precisely:

$$\iota_0^* h^*\omega = f^*\omega,$$
$$\iota_1^* h^*\omega = g^*\omega.$$

Now that we've done everything obvious, we have to interrupt our confident transcription of the proof for a moment so we can look for an *idea* for constructing α.

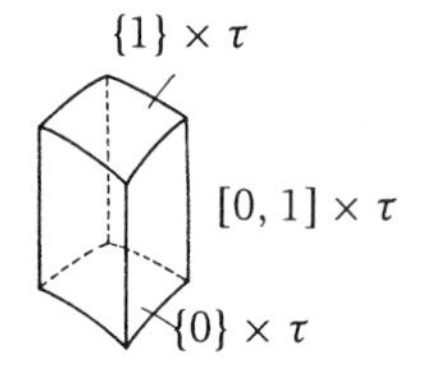

Figure 11.2. The prism over the k-cell τ

The cocycle $h^*\omega$ on $[0, 1] \times M$ does at least establish some kind of link between $f^*\omega$ and $g^*\omega$. The vague idea of somehow using $h^*\omega$ to define the desired $\alpha \in \Omega^{k-1}M$ is probably obvious enough. Where else could we start? So we have to take a closer look at the relationship of $h^*\omega$ to $f^*\omega$ and $g^*\omega$. Let τ be an oriented k-cell in M; then $[0, 1] \times \tau \subset [0, 1] \times M$ is the cylinder or *prism* over τ. Like any cocycle, $h^*\omega$ must respond with zero to the oriented boundary of $[0, 1] \times \tau$:

$$\int_{\partial([0,1]\times\tau)} h^*\omega = \int_{[0,1]\times\tau} dh^*\omega = 0$$

because $dh^*\omega = 0$. But the boundary consists of the top, bottom, and sides, and the top and bottom have opposite orientations. Now h is given on the top by g and on the bottom by f, so

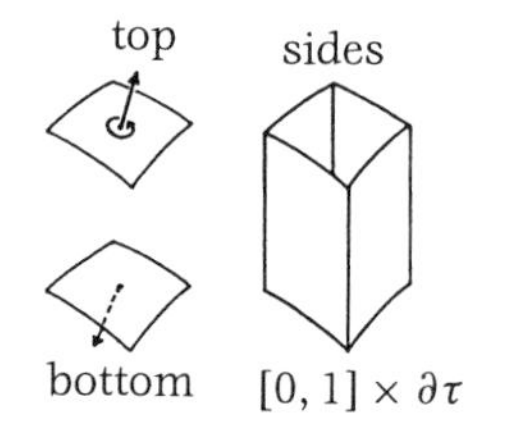

Figure 11.3. $\partial([0, 1]\times\tau) = \{1\}\times\tau\cup\{0\}\times\tau\cup[0, 1]\times\partial\tau$

$$\int_\tau g^*\omega - \int_\tau f^*\omega = \pm \int_{[0,1]\times\partial\tau} h^*\omega.$$

Of course, we could also figure out the sign by examining the orientations more carefully (the sign is positive), but that would be pedantic at this point. All that really matters is finding a $(k-1)$-form $\alpha \in \Omega^{k-1}M$ whose coboundary $d\alpha$ responds to τ as $h^*\omega$ does to $[0, 1] \times \partial\tau$. But since, in any case, $d\alpha$ responds to τ as α does to $\partial\tau$, we want

$$\int_\sigma \alpha = \int_{[0,1]\times\sigma} h^*\omega$$

for every oriented $(k-1)$-cell σ in M. In words: *α should respond to σ as $h^*\omega$ does to the prism over σ.* So if we just upgrade this from a requirement on α to the *definition* of α, we'll be done with the proof—at least intuitively.

But how can we make a precise definition out of the intuitive idea of a "prism operator"

$$P : \Omega^k([0,1] \times M) \longrightarrow \Omega^{k-1}M,$$

where we think of $P\eta$ as given by the effect of η on prisms? Well, the integral $\int_{[0,1]\times\sigma} \eta$ is defined as an ordinary multiple integral of the downstairs component function, and integrating with respect to the variable t gives

$$\int_{[0,1]\times\sigma} \eta = \int_{\sigma} \left(\int_0^1 \eta(\partial_t, \ldots)dt \right)$$

by Fubini. So all we have to do is set

$$P\eta(v_1, \ldots, v_{k-1}) := \int_0^1 \eta(\partial_t, v_1, \ldots, v_{k-1})dt$$

and we can be sure, on the basis of our reasoning, that one of the two k-forms

$$\alpha := \pm P h^* \omega$$

solves our problem.

11.4 Carrying Out the Proof

Now that we've found the idea, it will be easiest to carry out the proof just by verifying the desired property $d\alpha = g^*\omega - f^*\omega$, which in this case reads

$$dPh^*\omega = \iota_1^* h^*\omega - \iota_0^* h^*\omega.$$

If we consider an arbitrary η instead of the special $h^*\omega$, the geometric meaning of the operators d and P tells us that $Pd\eta$ will respond to an oriented k-cell τ in M as η does to the boundary of the prism over τ; since this boundary consists of bottom, top, and sides, we can thus expect (perhaps up to sign)

$$Pd\eta = (\iota_1^*\eta - \iota_0^*\eta) - dP\eta.$$

Notation. We denote by $v \lrcorner \eta$ the $(k-1)$-form $\eta(v, \dots)$ that results from inserting a vector v in the first slot of a k-form η.

Assertion. *For the **prism operator***

$$\begin{aligned} P : \Omega^k([0,1] \times M) &\longrightarrow \Omega^{k-1}M, \\ \eta &\longmapsto \int_0^1 (\partial_t \lrcorner \eta) dt, \end{aligned}$$

we have $Pd\eta = \iota_1^*\eta - \iota_0^*\eta - dP\eta$.

PROOF OF THE ASSERTION. The assertion is linear in η and local with respect to M, so it suffices to consider the two cases

(1) $\quad \eta = a\, dx^{\mu_1} \wedge \dots \wedge dx^{\mu_k}$ and

(2) $\quad \eta = b\, dt \wedge dx^{\mu_1} \wedge \dots \wedge dx^{\mu_{k-1}}$

in local coordinates on M.

CASE (1): Here $P\eta = 0$ since $\partial_t \lrcorner \eta = 0$, so certainly $dP\eta = 0$. Moreover,

$$d\eta = \dot{a}\, dt \wedge dx^{\mu_1} \wedge \dots \wedge dx^{\mu_k} + \sum_{i=1}^{n} \frac{\partial a}{\partial x^i} dx^i \wedge dx^{\mu_1} \wedge \dots \wedge dx^{\mu_k},$$

and hence

$$\begin{aligned} Pd\eta &= \left(\int_0^1 \dot{a}\, dt \right) dx^{\mu_1} \wedge \dots \wedge dx^{\mu_k} \\ &= (a(1,\cdot) - a(0,\cdot))\, dx^{\mu_1} \wedge \dots \wedge dx^{\mu_k} \\ &= \iota_1^*\eta - \iota_0^*\eta. \end{aligned} \qquad (1) \quad \square$$

CASE (2). Now $\iota_0^*\eta = \iota_1^*\eta = 0$ because $\iota_{t_0}^* dt = 0$ for any fixed t_0. So we must show that $Pd\eta = -dP\eta$. Since

$$d\eta = \sum_{i=1}^{n} \frac{\partial b}{\partial x^i} dx^i \wedge dt \wedge dx^{\mu_1} \wedge \ldots \wedge dx^{\mu_{k-1}},$$

it follows that

$$Pd\eta = -\sum_{i=1}^{n} \left(\int_0^1 \frac{\partial b}{\partial x^i} dt \right) dx^i \wedge dx^{\mu_1} \wedge \ldots \wedge dx^{\mu_{k-1}}.$$

On the other hand,

$$P\eta = \left(\int_0^1 b dt \right) dx^{\mu_1} \wedge \ldots \wedge dx^{\mu_{k-1}},$$

and hence

$$dP\eta = \sum_{i=1}^{n} \left(\int_0^1 \frac{\partial b}{\partial x^i} dt \right) dx^i \wedge dx^{\mu_1} \wedge \ldots \wedge dx^{\mu_{k-1}}. \qquad (2) \quad \square$$

The assertion has been proved. Now if ω is a k-cocycle on N and $\eta := h^*\omega$ the induced cocycle on $[0, 1] \times M$, then $d\eta = 0$, so $Pd\eta = 0$ and we obtain

$$g^*\omega - f^*\omega = \iota_1^* h^*\omega - \iota_0^* h^*\omega = \iota_1^*\eta - \iota_0^*\eta = dP\eta.$$

We have proved the following result.

Lemma. *If ω is a cocycle and h a homotopy between f and g, then the cocycles $g^*\omega$ and $f^*\omega$ differ only by the coboundary $d(Ph^*\omega)$.*

This completes the proof of the theorem on the homotopy invariance of de Rham cohomology.

11.5 The Poincaré Lemma

Now we reap a series of corollaries of homotopy invariance. The homotopies are always understood to be differentiable.

In fact, differentiable maps that are *continuously* homotopic are always differentiably homotopic, as a suitable approximation theorem shows. Since the homotopy class of any continuous map $f : M \to N$ must also contain differentiable representatives, de Rham cohomology is well defined and homotopy invariant even for the category of differentiable manifolds and *continuous* maps. But we won't go into this here.

Our first results follow from the fact that for $k > 0$, a k-form induced by a constant map must be zero (see Section 8.3).

Corollary 1. *If $f : M \to N$ is **null-homotopic** (i.e. homotopic to a constant map), then $f^* : H^kN \to H^kM$ is the zero map for all $k \geq 1$.*

Corollary 2. *If M is **contractible** (i.e. if $\mathrm{Id}_M : M \to M$ is null-homotopic) then $H^kM = 0$ for all $k \geq 1$.*

PROOF. $\mathrm{Id}_M^* : H^kM \to H^kM$ is the identity by the functorial property, but also zero by Corollary 1. □

Corollary 3. *On a contractible manifold, every positive-dimensional cocycle is a coboundary; in other words, if $\omega \in \Omega^kM$, $k > 0$, and $d\omega = 0$, then there exists $\alpha \in \Omega^{k-1}M$ with $d\alpha = \omega$.*

Corollary 4 (Poincaré lemma). *For an arbitrary manifold M, any positive-dimensional cocycle is locally a coboundary; that is, every point has an open neighborhood U in which, for every $\omega \in \Omega^kM$ with $k > 0$ and $d\omega = 0$, there exists $\alpha \in \Omega^{k-1}U$ with $d\alpha = \omega|U$.*

Any contractible open neighborhood U of p (any open "chart ball," for instance) will obviously work.

Another special case of Corollary 3 is also often called the Poincaré lemma, so we state it explicitly.

Corollary 5. *If $X \subset \mathbb{R}^n$ is open and star-shaped, then every positive-dimensional cocycle on X is a coboundary.*

This case also serves a special interest, for the following reason. Suppose we are explicitly given a "contraction" of a manifold M, a differentiable map

$$h : [0, 1] \times M \longrightarrow M \quad \text{with}$$
$$h_0 = \text{constant} \quad \text{and}$$
$$h_1 = \mathrm{Id}_M.$$

Then, by the lemma stated at the end of the proof of homotopy invariance, we also have an explicit integral formula that tells us, for any cocycle ω on M, *how* to find a form α with $d\alpha = \omega$. Imitating the term "antiderivative" used for functions, we might call α an "antiderivative form":

$$d\omega = 0 \Longrightarrow d(Ph^*\omega) = \omega,$$

so $\alpha = Ph^*\omega$ is an antiderivative of ω. Now, a domain $X \subset \mathbb{R}^n$ that is star-shaped with respect to $x_0 \in X$ has the simplest possible contraction, namely the straight-line contraction

$$h(t, x) := x_0 + t(x - x_0).$$

Thus we can also write down a completely explicit antiderivative for a cocycle $\omega \in \Omega^k X$. Without loss of generality, let $x_0 = 0$, so $h(t, x) = tx$ and

$$\omega = \sum_{\mu_1 < \cdots < \mu_k} \omega_{\mu_1 \ldots \mu_k} dx^{\mu_1} \wedge \ldots \wedge dx^{\mu_k}.$$

Then

$$h^*\omega_{\mu_1 \ldots \mu_k}(t, x) = \omega_{\mu_1 \ldots \mu_k}(tx),$$
$$h^*x^\mu = tx^\mu,$$
$$h^*dx^\mu = dh^*x^\mu = x^\mu dt + t dx^\mu$$

at each point $(t, x) \in [0, 1] \times X$. But $dt(\partial_t) = 1$ and $dx^\mu(\partial_t) = 0$ on $[0, 1] \times X$, so $\partial_t \lrcorner h^*\omega =$

$$\sum_{\mu_1 < \cdots < \mu_k} \sum_{i=1}^{k} (-1)^{i-1} t^{k-1} \omega_{\mu_1 \ldots \mu_k}(tx) x^{\mu_i} dx^{\mu_1} \wedge \ldots \widehat{\imath} \ldots \wedge dx^{\mu_k},$$

and since $Ph^*\omega$ was defined as $\int_0^1 \partial_t \lrcorner h^*\omega$ (see Section 11.4), we have the following result.

Corollary 6 ("Antiderivative formula for forms"). *Let $X \subset \mathbb{R}^n$ be open and star-shaped with respect to $x_0 = 0$, and let $\omega \in \Omega^k X$ be a cocycle (i.e. $d\omega = 0$). Then setting $\alpha :=$*

$$\sum_{\mu_1<\cdots<\mu_k} \sum_{i=1}^{k} (-1)^{i-1}\Big(\int_0^1 t^{k-1}\omega_{\mu_1\ldots\mu_k}(tx)dt\Big)x^{\mu_i}dx^{\mu_1}\wedge\ldots\widehat{\imath}\ldots\wedge dx^{\mu_k}$$

gives $d\alpha = \omega$.

Of course, we could check directly and mechanically that $d\alpha = \omega$. This would give a simple, elegant, and completely incomprehensible proof of the Poincaré lemma for star-shaped domains.

In Section 10.3 we saw how the three Cartan derivatives of the de Rham complex of an open subset of $\mathbb{R}^3$ correspond to the operators gradient, curl, and divergence. So translating the Poincaré lemma into classical vector analysis gives another corollary.

Corollary 7. *If $X \subset \mathbb{R}^3$ is open and contractible (star-shaped, for instance), then the following exist on X:*

(1) *for every vector field $\vec{a}$ with* $\operatorname{curl}\vec{a} = 0$, *a function f with* $\operatorname{grad} f = \vec{a}$;

(2) *for every vector field $\vec{b}$ with* $\operatorname{div}\vec{b} = 0$, *a vector field $\vec{a}$ with* $\operatorname{curl}\vec{a} = \vec{b}$;

(3) *for every function c, a vector field $\vec{b}$ with* $\operatorname{div}\vec{b} = c$.

11.6 The Hairy Ball Theorem

"*You can't comb the hair on a two-sphere.*" This vivid mnemonic has been adopted for the theorem that there is no nonvanishing vector field on an even-dimensional sphere.

It's hard to see at first what the problem has to do with the calculus of differential forms. Do we have to interpret the vector field as an $(n-1)$-form or something? Not at all—in fact, differential forms have absolutely nothing to do with it. The proof is a sample of *homological arguments*, and analogues work for other homology or cohomology theories.

Let M be an oriented *closed* n-dimensional manifold. The linear map

$$H^n M \longrightarrow \mathbb{R}, \quad [\omega] \longmapsto \int_M \omega,$$

is well defined by Stokes's theorem since $\int_M d\alpha = \int_\emptyset \alpha = 0$. In view of the homotopy invariance of $f^* : H^n N \to H^n M$, the composition of f^* with $\int_M$ is clearly homotopy invariant.

Corollary. *If M is an n-dimensional closed oriented manifold, then for all $f : M \to N$ the composition defined by*

$$H^n N \xrightarrow{f^*} H^n M \xrightarrow{\int_M} \mathbb{R}$$

is homotopy invariant.

We are about to derive the "hairy ball theorem" from this corollary, applied to $M = N = S^2$, and therefore point out that the corollary also follows easily from Stokes's theorem: Let h be a homotopy between f and g. Then

$$\int_M g^*\omega - \int_M f^*\omega = \int_{\partial([0,1]\times M)} h^*\omega = \int_{([0,1]\times M)} dh^*\omega = 0,$$

since $d\omega = 0$ and hence $dh^*\omega = h^* d\omega = 0$. Still, as a corollary of the homotopy invariance of de Rham cohomology the statement is logically in the right place. But now for the application.

Theorem. *Every differentiable vector field on an even-dimensional sphere has at least one zero.*

Proof. Let v be a nowhere-vanishing vector field on S^n, with n arbitrary for now. For any $x \in S^n$ we can think of $v(x)$ as a pointer toward the antipodal point $-x \in S^n$, and intuitively we see at once that the antipodal involution $\tau : S^n \to S^n$, $x \mapsto -x$, is homotopic to the identity. To check this formally, set

$$h(t, x) := \cos \pi t\, x + \sin \pi t \frac{v(x)}{\|v(x)\|}.$$

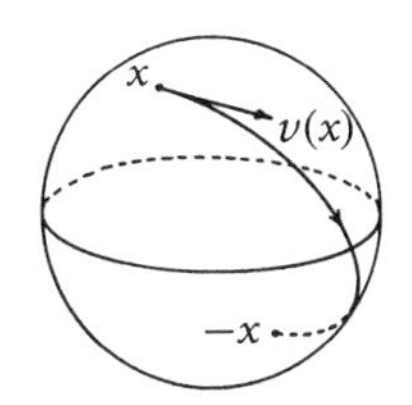

Figure 11.4. The vector $v(x)$ as a pointer

The homotopy invariance of the integral (a consequence of Stokes's theorem, as described above) implies that

$$\int_{S^n} \tau^*\omega = \int_{S^n} \omega$$

for all $\omega \in \Omega^n S^n$. (The homotopy invariance of de Rham cohomology even gives $\tau^*[\omega] = [\omega]$.) On the other hand, we know that

$$\int_{S^n} f^*\omega = \pm \int_{S^n} \omega$$

for every diffeomorphism $f : S^n \cong S^n$, where the sign depends on whether f preserves or reverses orientation (see Section 5.5); this is just the statement of the change-of-variables formula for integrals of differential forms. But the antipodal map $\tau : S^n \to S^n$ reverses orientation if and only if n is *even*. This can be seen, for example, as follows: For every $x \in S^n$, the differential of the diffeomorphism $-\mathrm{Id} : D^{n+1} \to D^{n+1}$ takes the outward normal $\vec{N}(x)$ at x to $\vec{N}(-x)$ at $-x$ (the differential is $-\mathrm{Id}_{\mathbb{R}^{n+1}}$ everywhere), so the diffeomorphism reverses the orientation of the boundary exactly when it reverses the overall orientation of D^{n+1}, and it obviously does the latter if and only if n is *even*. Thus

$$\int_{S^n} \tau^*\omega = -\int_{S^n} \omega$$

for all ω when n is even, and since there exist n-forms ω with $\int_{S^n} \omega \neq 0$ this contradicts homotopy invariance. Hence there can be no such vector field v for even n. □

This beautiful geometric theorem, provable in a number of ways, is not only interesting in itself but also a point of departure and a joint special case for various further developments (global properties of vector fields on manifolds, more generally of sections of vector bundles, Euler characteristic, characteristic classes, ...).

11.7 Test

(1) The cohomology class $[\eta] \subset \Omega^k M$ of a cocycle η of degree k is

☐ $\{\eta + \omega : \omega \in \Omega^k M,\ d\omega = 0\}$.

☐ $\{\eta + d\omega : \omega \in \Omega^{k-1} M\}$.

☐ $\{\eta + \omega : d\omega = d\eta\}$.

(2) What is meant by *anticommutativity* of the graded algebra H^*M is the following property of the wedge product: for all $[\omega] \in H^r M$ and $[\eta] \in H^s M$,

☐ $[\omega] \wedge [\eta] = -[\eta] \wedge [\omega]$.

☐ $[\omega] \wedge [\eta] = (-1)^{r+s} [\eta] \wedge [\omega]$.

☐ $[\omega] \wedge [\eta] = (-1)^{rs} [\eta] \wedge [\omega]$.

(3) $H^k M = 0$ if and only if

☐ for every $\omega \in \Omega^k M$ there exists $\eta \in \Omega^{k-1} M$ such that $d\eta = \omega$.

☐ for every $\omega \in \Omega^k M$ with $d\omega = 0$ there exists $\eta \in \Omega^{k-1} M$ such that $d\eta = \omega$.

☐ for every $\omega \in \Omega^k M$ of the form $\omega = d\eta$ we have $d\omega = 0$.

(4) The polar coordinates (r, φ) on $M := \mathbb{R}^2 \setminus \{0\}$ give a well-defined 1-form $d\varphi$. This 1-form is

☐ a cocycle, because $dd = 0$ is a local property.

☐ a coboundary, because $\varphi \in \Omega^0 M$.

☐ not a cocycle, and certainly not a coboundary, because φ can't be defined on all of $\mathbb{R}^2 \setminus \{0\}$ without a "jump."

(5) In connection with the prism operator, we had to consider $\partial_t \lrcorner\, \omega$. Now let x, y, and z denote the coordinates in $\mathbb{R}^3$. Then

☐ $\partial_x \lrcorner (dx \wedge dy + dy \wedge dz) = dy$.

☐ $\partial_x \lrcorner (dx \wedge dy + dy \wedge dz) = -dy$.

☐ $\partial_x \lrcorner (dx \wedge dy + dy \wedge dz) = dy \wedge dz$.

(6) Which of the following is true for the cylinder $M := S^1 \times \mathbb{R}$?

☐ $H^2M = 0$ because the fact that M is two-dimensional implies that $d : \Omega^2 M \to \Omega^3 M$ is zero.

☐ $H^2M = 0$ because M is the product of two one-dimensional manifolds.

☐ $H^2M = 0$ because the projection

$$\begin{array}{ccc} S^1 \times \mathbb{R} & \longrightarrow & S^1 \times \mathbb{R}, \\ (z, x) & \longmapsto & (z, 0) \end{array}$$

is homotopic to the identity and $H^2 S^1 = 0$.

(7) Let the maps $f, g : M \to N$ be homotopic. Is it true that $f^*\omega = g^*\omega$ for all cocycles $\omega \in \Omega^k N$?

☐ Yes, $f^* = g^*$ by the homotopy invariance theorem.

☐ No, because if $f^*\omega = g^*\omega$ for all cocycles, then $f = g$.

☐ No. On a contractible manifold, for example, the identity is homotopic to a constant map.

(8) Let p and q denote the north and south poles of the n-sphere S^n, $n > 1$. Then

☐ $S^n \setminus \{p\}$ is contractible because it is diffeomorphic to $\mathbb{R}^n$.

☐ S^n is not contractible because $H^n(S^n) \neq 0$.

☐ $S^n \setminus \{p, q\}$ is not contractible because the identity on S^{n-1} can be factored over $S^{n-1} \setminus \{p, q\}$:

$$S^{n-1} \longrightarrow S^n \setminus \{p, q\} \longrightarrow S^{n-1},$$

so $H^{n-1}(S^n \setminus \{p, q\}) \neq 0$.

(9) That *every* vector field with zero divergence on the open set $X \subset \mathbb{R}^3$ is the curl of a vector field on X is equivalent to

- □ X is connected (Poincaré lemma).
- □ $H^1X = 0$ (cocycles are coboundaries).
- □ $H^2X = 0$ (cocycles are coboundaries).

(10) Does an analogue of the hairy ball theorem hold for the even-dimensional real projective spaces $\mathbb{RP}^{2k}$?

- □ Yes, since $S^{2k} \to \mathbb{RP}^{2k}$ is a covering, and any vector field on $\mathbb{RP}^{2k}$ can be lifted to S^{2k}.
- □ No. For example, in homogeneous coordinates on the projective plane $\mathbb{RP}^2$, a nowhere-vanishing vector field is well defined by

$$[x_1 : x_2 : x_3] \longmapsto (x_1, x_2, x_3)/\|x\|.$$

- □ No. There are vector fields on S^{2k} that have zeros but do not vanish at a pair of antipodal points $\{\pm x\}$.

11.8 Exercises

EXERCISE 11.1. Prove that $H^1(S^2) = 0$.

EXERCISE 11.2. Prove directly from the definition that $[\omega] \to \int_{S^1} \omega$ defines an isomorphism $H^1(S^1) \cong \mathbb{R}$, and go on to show that $\dim H^1(S^1 \times S^1) \geq 2$.

EXERCISE 11.3. A map $f : M \to N$ is called a ***homotopy equivalence*** if it has a homotopy inverse, i.e. a map $g : N \to M$ such that $f \circ g$ and $g \circ f$ are homotopic to the identity maps of N and M, respectively. The manifolds or spaces M and N are then called ***homotopy equivalent***. Show that $\mathbb{R}^3 \setminus \{0\}$ and S^2 are homotopy equivalent but S^2 and $S^1 \times S^1$ are not.

11.9 Hints for the Exercises

FOR EXERCISE 11.1. For every closed 1-form (that is, every $\omega \in \Omega^1(S^2)$ with $d\omega = 0$) you have to find a function f with $df = \omega$. To do this, choose a point $q \in S^2$, say the south pole, and define

$$f(x) := \int_{\gamma} \omega =: \int_{q}^{x} \omega,$$

where γ denotes a path from q to x.

Don't think that I've already given away the solution: the real work is just starting! Why is this f even well defined? And why is $df = \omega$? The local solutions of the equation $df = \omega$ given by the Poincaré lemma are a great help in thinking about these questions.

That $H^1(M) = 0$ for any simply connected manifold can be proved in exactly the same way. But when $M = S^2$ life can be made a bit easier, for instance by applying the Poincaré lemma to the contractible subdomains $S^2 \setminus \{q\}$ and $S^2 \setminus \{p\}$ of S^2 and comparing the two functions to each other.

FOR EXERCISE 11.2. For the second part of the exercise you just need to use the functorial property of H^1. For example, consider the four maps given by projection onto and inclusion of the factors

$$S^1 \rightleftarrows S^1 \times S^1 \rightleftarrows S^1,$$

and apply the functor H^1 to them.

FOR EXERCISE 11.3. At first, when you're just getting to know the definitions, it's easier to answer yes than no to such topological existence questions—after all, if the thing in question exists you have some chance of finding and exhibiting it. But when that doesn't work, how can you be sure that it just *can't be done*? Later, though, the tables are turned, because you learn about functors that often yield nonexistence statements for free, while explicit constructions usually come at some cost in effort.

But in the first part of the present exercise, this cost in effort is modest, and a functor you can use for the second part is practically still on the table from the other two exercises.

Differential Forms on Riemannian Manifolds

12.1 Semi-Riemannian Manifolds

For a fuller development of the calculus of differential forms, we now proceed to Riemannian manifolds. Here we will encounter the star operator, the Laplace–de Rham operator, the Hodge decomposition, and Poincaré duality. We begin by considering the somewhat more general semi-Riemannian manifolds.

Before introducing Riemannian and semi-Riemannian manifolds, I would like to remind you of a few linear-algebraic concepts and facts: A symmetric bilinear form $\langle \cdot\,, \cdot \rangle$ on an n-dimensional real vector space V is called ***non-degenerate*** if

$$\begin{array}{rcl} V & \longrightarrow & V^*, \\ v & \longmapsto & \langle v, \cdot \rangle \end{array}$$

is an isomorphism, and this occurs if and only if the $n \times n$ matrix G given by

$$g_{\mu\nu} := \langle v_\mu, v_\nu \rangle$$

has full rank for some (hence any) basis $(v_1, \ldots, v_n)$ of V. A basis can be chosen in such a way that G has the form

$$\begin{pmatrix} +1 & & & & & \\ & \ddots & & & & \\ & & +1 & & & \\ & & & -1 & & \\ & & & & \ddots & \\ & & & & & -1 \end{pmatrix} \begin{matrix} \left.\vphantom{\begin{matrix}1\\1\\1\end{matrix}}\right\} r \\ \left.\vphantom{\begin{matrix}1\\1\\1\end{matrix}}\right\} s \end{matrix}$$

The number s of -1's on the diagonal is independent of the choice of such a basis (Sylvester's law of inertia), and is called the ***index*** of the symmetric bilinear form. The quadratic form $q : V \to \mathbb{R}$ corresponding to $\langle \cdot , \cdot \rangle$ is defined by

$$q(v) := \langle v, v \rangle,$$

and we can recover $\langle \cdot , \cdot \rangle$ from q through the identity

$$\langle v, w \rangle = \frac{1}{2}(q(v + w) - q(v) - q(w)).$$

The pair (V, q) or $(V, \langle \cdot , \cdot \rangle)$ is called a nondegenerate ***quadratic space*** of index s, and a ***Euclidean space*** in the positive definite case $s = 0$.

Definition. By a ***semi-Riemannian manifold of index*** s, we mean a pair $(M, \langle \cdot , \cdot \rangle)$ consisting of a manifold M and a family

$$\langle \cdot , \cdot \rangle = \{\langle \cdot, \cdot \rangle_p\}_{p \in M}$$

of symmetric bilinear forms $\langle \cdot , \cdot \rangle_p$ of index s on T_pM that is differentiable in the following obvious sense: for the charts (U, h) of some (hence every) atlas on M, the functions $g_{\mu\nu} : U \to \mathbb{R}$ defined by $p \mapsto \langle \partial_\mu, \partial_\nu \rangle_p$ are differentiable. In the positive definite case $s = 0$, we call $(M, \langle \cdot , \cdot \rangle)$ a ***Riemannian manifold***.

The family $\langle \cdot , \cdot \rangle$ is called the Riemannian or semi-Riemannian ***metric*** of $(M, \langle \cdot , \cdot \rangle)$. We retain the "$p$" in the notation $\langle\cdot\,,\cdot\rangle_p$ only when clarity seems to demand it, and otherwise write $\langle v, w\rangle_p =: \langle v, w\rangle$ for $v, w \in T_pM$.

Submanifolds of $\mathbb{R}^n$ are Riemannian manifolds in a canonical way. But an arbitrary manifold, with charts (U_λ, h_λ), can also be provided with a Riemannian metric $\langle\cdot\,,\cdot\rangle$ by choosing a partition of unity $\{\tau_\lambda\}_{\lambda\in\Lambda}$ such that $\operatorname{supp}\tau_\lambda \subset U_\lambda$ and setting

$$\langle v, w\rangle_p := \sum_{\lambda\in\Lambda} \tau_\lambda(p)\,\langle v, w\rangle_\lambda,$$

where $\langle v, w\rangle_\lambda$ denotes the Riemannian metric transferred from $U'_\lambda \subset \mathbb{R}^n$ to U_λ by dh_λ. Observe, though, that the same procedure fails in general if we try to apply it to construct a semi-Riemannian metric of index $0 < s < n$. We could start with the semi-Riemannian metric

$$\langle x, y\rangle_{n-s,s} := \sum_{\mu=1}^{n-s} x^\mu y^\mu - \sum_{\nu=n-s+1}^{n} x^\nu y^\nu$$

on $\mathbb{R}^n$ and write down a formula analogous to the one above for $\langle\cdot\,,\cdot\rangle$ on M. But since the property of being nondegenerate and of index s, in contrast to positive definiteness, is not convex (see, for instance, [J:*Top*], p. 120), it does not carry over in general from $\langle\cdot\,,\cdot\rangle_\lambda$ to the convex combination $\sum_\lambda \tau_\lambda(p)\langle\cdot\,,\cdot\rangle_\lambda$. On the even-dimensional spheres S^n, for example, there is no semi-Riemannian metric of index 1 (or $n-1$), as can be shown using the hairy ball theorem and a covering argument ([J: *Top*], pp. 152–153).

Semi-Riemannian n-dimensional manifolds ($n \geq 2$) of index 1 or $n-1$ are called ***Lorentz manifolds***. Changing the sign of the metric interchanges these indices. We follow the convention of taking the index of Lorentz manifolds to be $n-1$. Real space-time, via a metric $\langle\cdot\,,\cdot\rangle$ given physically, is a four-dimensional Lorentz manifold. This circumstance was historical and is still a principal motive for extending Riemannian geometry to semi-Riemannian manifolds. In the general theory of relativity, the differential geometry of Lorentz manifolds plays an important role both conceptually and technically, and in particle physics the Lorentz metric, through the theory of special relativity, is ubiquitous.

Our first goal will be to define the star operator

$$* : \Omega^k M \xrightarrow{\cong} \Omega^{n-k} M$$

for an oriented n-dimensional semi-Riemannian manifold M. This is done for each individual p by means of a star operator

$$* : \mathrm{Alt}^k T_p M \xrightarrow{\cong} \mathrm{Alt}^{n-k} T_p M,$$

so we set manifolds aside for the time being and return yet again to linear algebra.

12.2 The Scalar Product of Alternating k-Forms

We begin with a linear-algebraic observation about finite-dimensional real vector spaces that involves no additional structures such as orientation or metric: $(\mathrm{Alt}^k V)^*$ and $\mathrm{Alt}^k(V^*)$ are canonically the same. The following lemma makes this more precise.

Lemma. *If we interpret each linear form*

$$\varphi \in \mathrm{Hom}(\mathrm{Alt}^k V, \mathbb{R}) = (\mathrm{Alt}^k V)^*$$

on $\mathrm{Alt}^k V$ *as an alternating k-form (denoted by* $\widetilde{\varphi}$*) on* V^* *by setting*

$$\widetilde{\varphi}(\alpha^1, \ldots, \alpha^k) := \varphi(\alpha^1 \wedge \ldots \wedge \alpha^k)$$

for any $\alpha^1, \ldots, \alpha^k \in V^*$*, we obtain an equivalence of the two functors* $(\mathrm{Alt}^k -)^*$ *and* $\mathrm{Alt}^k(-^*)$ *from the category of finite-dimensional real vector spaces and linear maps to itself. In other words, for every linear map* $f : V \to W$ *between finite-*

dimensional real vector spaces, the following diagram is commutative:

$$\begin{array}{ccc} (\mathrm{Alt}^k V)^* & \xrightarrow{\cong} & \mathrm{Alt}^k(V^*) \\ {\scriptstyle (\mathrm{Alt}^k f)^*}\big\downarrow & & \big\downarrow{\scriptstyle \mathrm{Alt}^k(f^*)} \\ (\mathrm{Alt}^k W)^* & \xrightarrow{\cong} & \mathrm{Alt}^k(W^*) \end{array}$$

PROOF. The spaces $(\mathrm{Alt}^k V)^*$ and $\mathrm{Alt}^k(V^*)$ have the same dimension. Moreover, if $\varphi(\alpha^1 \wedge \ldots \wedge \alpha^k) = 0$ for all $\alpha^1, \ldots, \alpha^k \in V^*$, then $\varphi = 0 \in (\mathrm{Alt}^k V)^*$. So the canonical map is injective and hence an isomorphism. Its compatibility with f follows from the naturality of the wedge product. □

Now let a nondegenerate symmetric bilinear form $\langle \cdot , \cdot \rangle$ be given on V. This is also called a (not necessarily positive definite) ***scalar product***. We adopt the following suggestive notation from [AM].

Notation. If $\langle \cdot , \cdot \rangle$ is a nondegenerate bilinear form on a finite-dimensional real vector space V, we denote the isomorphism between V and V^* given by $v \mapsto \langle v, \cdot \rangle$ and its inverse by

$$V \underset{\sharp}{\overset{\flat}{\rightleftarrows}} V^*.$$

Instead of $\flat(v)$ we sometimes write ${}^\flat v$ or $v^\flat$, whichever is most convenient, and similarly for $\sharp$.

The meaning of the notation can be inferred from the symbols $\sharp$ and $\flat$ in music, which, as you know, are read "sharp" and "flat." The linear form α is "sharpened" to the vector ${}^\sharp\alpha$ by $\sharp$.

By the lemma above, an isomorphism $V \cong V^*$ also induces an isomorphism $\mathrm{Alt}^k V \cong (\mathrm{Alt}^k V)^*$ and thus a bilinear form on $\mathrm{Alt}^k V$, or more precisely:

Defining Lemma (Scalar product on the space of forms). *If $(V, \langle \cdot , \cdot \rangle)$ is an n-dimensional nondegenerate quad-*

ratic space, then on $\mathrm{Alt}^k V$ *there is a canonical bilinear form* $\langle\cdot,\cdot\rangle$ *that is also symmetric and nondegenerate, namely the one given by*

$$(\mathrm{Alt}^k V)^* \xrightarrow[\text{canon}]{\cong} \mathrm{Alt}^k V^* \xrightarrow[\mathrm{Alt}^k \flat]{\cong} \mathrm{Alt}^k V.$$

PROOF. Let $\omega, \eta \in \mathrm{Alt}^k V$ and let $\varphi, \psi \in (\mathrm{Alt}^k V)^*$ be their preimages under the map above. All we have to prove is the symmetry condition

$$\langle \omega, \eta \rangle := \varphi(\eta) = \psi(\omega) =: \langle \eta, \omega \rangle.$$

If we trace how ω results from

$$\varphi \longmapsto \widetilde{\varphi} \longmapsto \omega,$$

we find that

$$\omega(v_1, \ldots, v_k) = \widetilde{\varphi}({}^\flat v_1, \ldots, {}^\flat v_k) = \varphi({}^\flat v_1 \wedge \ldots \wedge {}^\flat v_k),$$

and similarly for η and ψ. Now let $(e_1, \ldots, e_n)$ be an orthonormal basis, abbreviated ***o.n. basis***, of the quadratic space V; that is, $\langle e_\mu, e_\nu \rangle = \pm\delta_{\mu\nu}$. We write $\langle e_\mu, e_\mu \rangle =: \varepsilon_\mu = \pm 1$. Let $(\delta^1, \ldots, \delta^n)$ be the corresponding dual basis of V^*. Observe that

$${}^\flat e_\mu = \varepsilon_\mu \delta^\mu$$

for every μ (no summation), since

$${}^\flat e_\mu(e_\nu) := \langle e_\mu, e_\nu \rangle = \varepsilon_\mu \delta_{\mu\nu} = \varepsilon_\mu \delta^\mu(e_\nu)$$

for every ν.

Without loss of generality, we now set

$$\omega = \delta^{\mu_1} \wedge \ldots \wedge \delta^{\mu_k},$$
$$\eta = \delta^{\nu_1} \wedge \ldots \wedge \delta^{\nu_k}.$$

Then

$$\begin{aligned}
\langle \omega, \eta \rangle &= \varphi(\delta^{\nu_1} \wedge \ldots \wedge \delta^{\nu_k}) \\
&= \varepsilon_{\nu_1} \cdot \ldots \cdot \varepsilon_{\nu_k} \varphi({}^\flat e_{\nu_1} \wedge \ldots \wedge {}^\flat e_{\nu_k}) \\
&= \varepsilon_{\nu_1} \cdot \ldots \cdot \varepsilon_{\nu_k} \omega(e_{\nu_1}, \ldots, e_{\nu_k}) \\
&= \varepsilon_{\nu_1} \cdot \ldots \cdot \varepsilon_{\nu_k} \delta^{\mu_1} \wedge \ldots \wedge \delta^{\mu_k}(e_{\nu_1}, \ldots, e_{\nu_k}).
\end{aligned}$$

Hence $\langle\omega, \eta\rangle = \varepsilon_{\nu_1} \cdot \ldots \cdot \varepsilon_{\nu_k} \operatorname{sgn}\tau$ if the $\mu_1, \ldots, \mu_k$ are distinct and come from $\nu_1, \ldots, \nu_k$ through a permutation τ of the indices $1, \ldots, k$; otherwise $\langle\omega, \eta\rangle$ is zero. In particular, $\langle\omega, \eta\rangle = \langle\eta, \omega\rangle$. □

This proof also gives us a formula for computing the scalar product on $\mathrm{Alt}^k V$.

Lemma (Orthonormal basis in the space of forms). *If $(e_1, \ldots, e_n)$ is an o.n. basis of the quadratic space V and $(\delta^1, \ldots, \delta^n)$ denotes the dual basis, then*

$$(\delta^{\mu_1} \wedge \ldots \wedge \delta^{\mu_k})_{\mu_1 < \cdots < \mu_k}$$

is an o.n. basis of $\mathrm{Alt}^k V$ and

$$\langle \delta^{\mu_1} \wedge \ldots \wedge \delta^{\mu_k}, \delta^{\mu_1} \wedge \ldots \wedge \delta^{\mu_k} \rangle = \varepsilon_{\mu_1} \cdot \ldots \cdot \varepsilon_{\mu_k},$$

where $\varepsilon_\mu := \langle e_\mu, e_\mu \rangle$.

12.3 The Star Operator

Now we add an orientation to our data. To begin with, we have a canonical "volume form" $\omega_V \in \mathrm{Alt}^n V$.

Defining Lemma (Volume form). *Let V be an n-dimensional oriented nondegenerate quadratic space. The alternating n-form $\omega_V \in \mathrm{Alt}^n V$ that assigns the value $+1$ to some (hence every) positively oriented o.n. basis is called the* ***volume form*** *of V.*

Proof of the assertion ("hence every"). Let $(e_1', \ldots, e_n')$ be a second positively oriented o.n. basis and let $f : V \to V$ be the linear transformation with $f(e_\mu) = e_\mu'$. Then

$$\omega(e_1', \ldots, e_n') = f^*\omega(e_1, \ldots, e_n) = \det f$$

by the lemma in Section 3.3. So we must show that $\det f = +1$. If A is the matrix of f with respect to $(e_1, \ldots, e_n)$, then $e_i' = \sum a_{ji} e_j$ and hence $\langle e_i', e_k' \rangle = \sum\sum a_{ji} a_{lk} \langle e_j, e_l \rangle$, or in matrix notation

$$G' = {}^t\!A \cdot G \cdot A.$$

Since $|\det G| = |\det G'| = 1$ (o.n. property of the bases), this implies first that $|\det A| = 1$ and then, since f is orientation-preserving, that $\det f = \det A = +1$. □

Incidentally, if the second basis is not necessarily orthonormal but just positively oriented, the same calculation shows that

$$\det f = \sqrt{|\det G'|}.$$

Writing this in a frequently used notation gives the following result.

Lemma (Volume form formula). *Let V be an oriented n-dimensional nondegenerate quadratic space, $(v_1, \ldots, v_n)$ a positively oriented basis, and $(\delta^1, \ldots, \delta^n)$ the dual basis. Then the volume form is*

$$\omega_V = \sqrt{|g|}\, \delta^1 \wedge \ldots \wedge \delta^n,$$

where g is the determinant of the $n \times n$ matrix given by

$$g_{\mu\nu} := \langle v_\mu, v_\nu \rangle.$$

Now we define the star operator.

Defining Lemma (Star operator). *If V is an oriented n-dimensional nondegenerate quadratic space and $\omega_V \in \mathrm{Alt}^n V$ its canonical volume form, then for every k there is exactly one linear map*

$$* : \mathrm{Alt}^k V \longrightarrow \mathrm{Alt}^{n-k} V$$

(the ***star operator****) such that*

$$\eta \wedge *\zeta = \langle \eta, \zeta \rangle \omega_V$$

for all $\eta, \zeta \in \mathrm{Alt}^k V$.

Proof. We prove uniqueness first. Let $(\delta^1, \ldots, \delta^n)$ be the dual of a positively oriented basis and let $\lambda_1 < \cdots < \lambda_k$ and $\mu_1 < \cdots < \mu_k$ be ordered indices. By the lemma on o.n. bases in the space of forms (Section 12.2), the requirement also implies that

$$\begin{aligned}&\delta^{\lambda_1} \wedge \ldots \wedge \delta^{\lambda_k} \wedge *(\delta^{\mu_1} \wedge \ldots \wedge \delta^{\mu_k})\\&= \begin{cases} \varepsilon_{\mu_1} \cdot \ldots \cdot \varepsilon_{\mu_k} \omega_V & \text{if } \mu_i = \lambda_i \text{ for } i = 1, \ldots, k, \\ 0 & \text{otherwise.} \end{cases}\end{aligned}$$

But this means that the sum on the right-hand side of the equation

$$*(\delta^{\mu_1} \wedge \ldots \wedge \delta^{\mu_k}) = \sum_{\nu_1 < \cdots < \nu_{n-k}} a_{\nu_1 \ldots \nu_{n-k}} \delta^{\nu_1} \wedge \ldots \wedge \delta^{\nu_{n-k}}$$

can have only one nonzero component, which must correspond to the complementary multi-index

$$\nu_1 < \cdots < \nu_{n-k}.$$

More precisely,

$$*(\delta^{\mu_1} \wedge \ldots \wedge \delta^{\mu_k}) = \varepsilon_{\mu_1} \cdot \ldots \cdot \varepsilon_{\mu_k} \operatorname{sgn} \tau \cdot \delta^{\nu_1} \wedge \ldots \wedge \delta^{\nu_{n-k}},$$

where $\nu_1 < \cdots < \nu_{n-k}$ is complementary to $\mu_1 < \cdots < \mu_k$ and where τ denotes the permutation that takes $(1, \ldots, n)$ to $(\mu_1, \ldots, \mu_k, \nu_1, \ldots, \nu_{n-k})$. In particular, $*$ is uniquely determined by this necessary condition.

Conversely, given a fixed positively oriented o.n. basis, we use this formula to define $*$. Then the requirement that $\eta \wedge *\zeta = \langle \eta, \zeta \rangle \omega_V$ is satisfied if η, ζ are basis elements of $\mathrm{Alt}^k V$, and this suffices since the requirement is bilinear. This proves the existence of the star operator. □

The formula above also holds without the conditions $\mu_1 < \cdots < \mu_k$ and $\nu_1 < \cdots < \nu_{n-k}$ because $\operatorname{sgn} \tau$ captures the sign changes caused by permuting the indices. We state this more formally as follows.

Note 1. *For any positively oriented o.n. basis and any permutation τ,*

$$*(\delta^{\tau(1)} \wedge \ldots \wedge \delta^{\tau(k)}) = \varepsilon_{\tau(1)} \cdot \ldots \cdot \varepsilon_{\tau(k)} \operatorname{sgn} \tau \cdot \delta^{\tau(k+1)} \wedge \ldots \wedge \delta^{\tau(n)}.$$

This also implies that, up to sign, $*\eta$ responds to vectors that form part of an orthonormal basis as η does to the complementary or remaining vectors:

Note 2. *If* $(e_1, \ldots, e_n)$ *is a positively oriented o.n. basis, then*

$$*\eta(e_{\tau(k+1)}, \ldots, e_{\tau(n)}) = \varepsilon_{\tau(1)} \cdot \ldots \cdot \varepsilon_{\tau(k)} \operatorname{sgn} \tau \cdot \eta(e_{\tau(1)}, \ldots, e_{\tau(k)})$$

for every $\eta \in \mathrm{Alt}^k V$ *and every permutation* τ. *In particular,* $*1 = \omega_V$ and $*\omega_V = (-1)^{\mathrm{index}\, V} 1$.

Hence the composition

$$\mathrm{Alt}^k V \xrightarrow{\;*\;} \mathrm{Alt}^{n-k} V \xrightarrow{\;*\;} \mathrm{Alt}^k V$$

is the identity, up to sign, and since the index of V counts the number of times -1 occurs as a factor in $\varepsilon_1 \cdot \ldots \cdot \varepsilon_n$, this sign is given by the following formula:

Note 3. $** = (-1)^{k(n-k)+\mathrm{index}\, V} \mathrm{Id}_{\mathrm{Alt}^k V}$.

We began by setting $\eta \wedge *\zeta = \langle \eta, \zeta \rangle \omega_V$ as the characterizing property of the star operator. Now that we know the sign of $**$, we can read off more information from this definition.

Note 4. $\eta \wedge \zeta = (-1)^{k(n-k)+\mathrm{index}\, V} \langle \eta, *\zeta \rangle \omega_V$ *for all* $\eta \in \mathrm{Alt}^k V$ *and all* $\zeta \in \mathrm{Alt}^{n-k} V$.

Our next result follows from Note 4 and the definition of $*$.

Note 5. $\langle *\eta, *\zeta \rangle = (-1)^{\mathrm{index}\, V} \langle \eta, \zeta \rangle$ *for all* η *and* ζ *in* $\mathrm{Alt}^k V$.

Finally, we also mention that the star operator, as follows directly from its definition, changes sign under a change of orientation because the canonical volume form changes sign while the scalar product remains unchanged.

Now we turn from the star operator on a single vector space to manifolds and their tangent spaces. Let M be an n-dimensional oriented semi-Riemannian manifold. For each individual tangent space T_pM and each k, the three defining lemmas in Sections 12.2 and 12.3 give a scalar product $\langle \cdot\,, \cdot \rangle_p$ on $\mathrm{Alt}^k T_pM$, a canonical volume form $\omega_{T_pM} \in \mathrm{Alt}^n T_pM$, and a star operator

$$* : \mathrm{Alt}^k T_pM \longrightarrow \mathrm{Alt}^{n-k} T_pM,$$

which all depend differentiably on p.

Definition. Let M be an n-dimensional oriented semi-Riemannian manifold. Then the ***canonical volume form*** $\omega_M \in \Omega^n M$, the ***scalar product***

$$\langle \cdot , \cdot \rangle : \Omega^k M \times \Omega^k M \to C^\infty(M)$$

of k-forms, and the ***star operator***

$$* : \Omega^k M \longrightarrow \Omega^{n-k} M$$

are defined in the obvious way via the corresponding objects on the tangent spaces.

12.4 The Coderivative

The star operator translates the de Rham complex, in which the degrees of the differential forms increase, into an equivalent complex in which they decrease:

$$\begin{array}{ccccccccccc}
0 \longrightarrow & \Omega^0 M & \xrightarrow{d} & \Omega^1 M & \xrightarrow{d} \cdots \xrightarrow{d} & \Omega^{n-1} M & \xrightarrow{d} & \Omega^n M & \longrightarrow 0 \\
& \cong \big\downarrow * & & \cong \big\downarrow * & & \cong \big\downarrow * & & \cong \big\downarrow * & \\
0 \longrightarrow & \Omega^n M & \longrightarrow & \Omega^{n-1} M & \longrightarrow \cdots \longrightarrow & \Omega^1 M & \longrightarrow & \Omega^0 M & \longrightarrow 0
\end{array}$$

The Cartan derivative d goes to $* \, d \, *^{-1}$, and up to sign this is the coderivative δ. But the sign is subject to nonuniform conventions; we fix one as follows.

Definition. The ***coderivative***

$$\delta : \Omega^{n-k} M \longrightarrow \Omega^{n-k-1} M$$

on an n-dimensional semi-Riemannian manifold M is defined by

$$\delta := (-1)^k * d *^{-1} .$$

The coderivative is obviously independent of the orientation of M. The meaning of the sign becomes clear when we consider the *formal adjoint*, or *dual*, operator d' of d with respect to the scalar product. What this means is the following: Taking the pointwise scalar product of k-forms η, ζ on a manifold M defines a function $\langle \eta, \zeta \rangle \in C^\infty(M)$, and integrating over M by means of the volume form gives us a number, which we denote by $\langle\langle \eta, \zeta \rangle\rangle \in \mathbb{R}$ to make the distinction clear. We make this more precise.

Notation. For k-forms $\eta, \zeta \in \Omega^k M$ whose supports have compact intersection, we set

$$\langle\langle \eta, \zeta \rangle\rangle := \int_M \langle \eta, \zeta \rangle \omega_M = \int_M \eta \wedge *\zeta.$$

The differential operator d' dual to $d : \Omega^k M \to \Omega^{k+1} M$ should satisfy

$$\langle\langle d\eta, \zeta \rangle\rangle = \langle\langle \eta, d'\zeta \rangle\rangle$$

for all $\eta \in \Omega^k M$ and $\zeta \in \Omega^{k+1} M$ with compact support in $M \setminus \partial M$. In particular, it should be an operator from $\Omega^{k+1} M$ to $\Omega^k M$. By the product rule,

$$d(\eta \wedge *\zeta) = d\eta \wedge *\zeta + (-1)^k \eta \wedge d * \zeta.$$

Since we know the sign of $**$ by Note 3 in 12.3, we can easily convert $d * \zeta$ to $\pm * (*d*^{-1})\zeta$. Hence

$$\begin{aligned} d * \zeta &= (-1)^{(n-k)k + \operatorname{index} M} * *d * \zeta \\ &= (-1)^{(n-k)k + \operatorname{index} M} * *d * * *^{-1} \zeta \\ &= (-1)^{(n-k)k + (n-k-1)(k+1)} * *d *^{-1} \zeta \\ &= (-1)^{n-1} * (*d*^{-1})\zeta = (-1)^k * \delta\zeta \end{aligned}$$

for $\zeta \in \Omega^{k+1} M$. Our choice of sign for δ thus gives the following product rule.

Lemma. *For all $\eta \in \Omega^k M$ and $\zeta \in \Omega^{k+1} M$,*

$$d(\eta \wedge *\zeta) = d\eta \wedge *\zeta + \eta \wedge *\delta\zeta.$$

If the intersection of the supports of η and ζ is also compact in $M \setminus \partial M$, then $\int_M d(\eta \wedge *\zeta) = 0$ by Stokes's theorem, and we have the following corollary.

Corollary (Duality formula for the coderivative).

$$\langle\langle d\eta, \zeta \rangle\rangle + \langle\langle \eta, \delta\zeta \rangle\rangle = 0$$

for $\eta \in \Omega^k M$, $\zeta \in \Omega^{k+1} M$ whose supports have compact intersection in $M \setminus \partial M$.

Thus each $-\delta$ is dual to d by our sign convention for the coderivative. The opposite convention, which makes δ and d dual to each other, is also used (see, for instance, [W]).

Up to this point we have denoted all the operators in the de Rham complex by the same symbol d. Now we want to include the index k in the notation.

Notation. When necessary, the Cartan derivative and coderivative will be denoted by the more precise symbols d_k and δ_k, as follows:

$$\begin{array}{ccc} \Omega^k M & \xrightarrow{d_k} & \Omega^{k+1} M \\ {\scriptstyle *}\big\downarrow{\scriptstyle \cong} & & {\scriptstyle *}\big\downarrow{\scriptstyle \cong} \\ \Omega^{n-k} M & \xrightarrow{(-1)^k \delta_k} & \Omega^{n-k-1} M \end{array}$$

Thus $(-1)^k \delta_k$ is *conjugate* to d_k by means of $*$, and by the duality formula above, $-\delta_k$ is dual to (or *the formal adjoint of*) d_{n-k-1}. The double meaning of the coderivative as (up to sign) both conjugate and adjoint to the Cartan derivative establishes a relationship between d_k and d_{n-k-1}, which we will now examine more closely.

12.5 Harmonic Forms and the Hodge Theorem

In what follows, let M be an n-dimensional oriented *compact Riemannian manifold without boundary*. The scalar product $\langle\langle \cdot\,, \cdot \rangle\rangle$ is defined on all of $\Omega^k M$ by compactness, and because $\partial M = \emptyset$ the duality formula

$$\langle\langle d\eta, \zeta \rangle\rangle + \langle\langle \eta, \delta\zeta \rangle\rangle = 0$$

for the coderivative holds for *all* $\eta \in \Omega^k M$ and $\zeta \in \Omega^{k+1}M$. Finally, since the scalar product on M is now assumed to be positive definite, the scalar products $\langle \cdot , \cdot \rangle$ on the individual spaces $\text{Alt}^k T_p M$ and $\langle\langle \cdot , \cdot \rangle\rangle$ on $\Omega^k M$ are also positive definite, and this turns the $\Omega^k M$ into Euclidean vector spaces.

We now consider a portion of the sequences of the Cartan derivatives and coderivatives:

$$\Omega^{k-1}M \underset{\delta}{\overset{d}{\rightleftarrows}} \Omega^k M \underset{\delta}{\overset{d}{\rightleftarrows}} \Omega^{k+1}M,$$

or more precisely

$$\Omega^{k-1}M \underset{\delta_{n-k}}{\overset{d_{k-1}}{\rightleftarrows}} \Omega^k M \underset{\delta_{n-k-1}}{\overset{d_k}{\rightleftarrows}} \Omega^{k+1}M.$$

In the Euclidean space $(\Omega^k M, \langle\langle \cdot , \cdot \rangle\rangle)$, the operators are adjoints of each other. So it is trivial that

$$\ker d_k = (\operatorname{im} \delta_{n-k-1})^{\perp}$$

and

$$\ker \delta_{n-k} = (\operatorname{im} d_{k-1})^{\perp}.$$

The first equation holds because $d\eta = 0 \iff \langle\langle d\eta, \zeta \rangle\rangle = 0$ for all $\zeta \iff \langle\langle \eta, \delta\zeta \rangle\rangle = 0$ for all $\zeta \iff \eta \in (\operatorname{im} \delta)^{\perp}$, and similarly for $\ker \delta$.

For vector subspaces $V_0 \subset V$ of *finite-dimensional* Euclidean spaces V, we always have $V = V_0 \oplus V_0^{\perp}$. So if we were permitted to think of $\Omega^k M$ as finite-dimensional, we could conclude that

$$\Omega^k M = \ker d \oplus \operatorname{im} \delta = \ker \delta \oplus \operatorname{im} d.$$

The conclusion is actually true, but although this decomposition of $\Omega^k M$ seems to be within easy reach, the proof requires methods from the theory of elliptic differential operators and is beyond the scope of this course. See Chapter 6 of [W], for example.

Theorem (here without proof). *If M is an oriented n-dimensional closed Riemannian manifold, then*

$$\Omega^k M = \ker d_k \oplus \operatorname{im} \delta_{n-k-1} = \ker \delta_{n-k} \oplus \operatorname{im} d_{k-1}$$

as an orthogonal direct sum with respect to the scalar product on $\Omega^k M$ defined by

$$\langle\langle \eta, \zeta \rangle\rangle := \int_M \eta \wedge *\zeta.$$

This theorem, which looks a bit technical at first, lies at the heart of Hodge theory for the de Rham complex. Our first corollary is the following.

Corollary. *For M as above,*

$$\begin{aligned} \ker d_k &= \operatorname{im} d_{k-1} \oplus (\ker d_k \cap \ker \delta_{n-k}), \\ \ker \delta_{n-k} &= \operatorname{im} \delta_{n-k-1} \oplus (\ker d_k \cap \ker \delta_{n-k}). \end{aligned}$$

The k-forms $\eta \in \Omega^k M$ making their appearance here—those for which $d\eta = 0$ *and* $\delta\eta = 0$—belong to the kernel of the ***Laplace-de Rham***, or ***Laplace-Beltrami, operator***

$$\Delta := d\delta + \delta d : \Omega^k M \longrightarrow \Omega^k M.$$

The k-forms satisfying $\Delta\eta = 0$ are called ***harmonic forms***. For oriented closed Riemannian manifolds, the duality formula for the coderivative (see Section 12.4) implies that

$$\langle\langle \Delta\eta, \eta \rangle\rangle = -\langle\langle \delta\eta, \delta\eta \rangle\rangle - \langle\langle d\eta, d\eta \rangle\rangle$$

for all $\eta \in \Omega^k M$, so if $\Delta\eta = 0$, then $\delta\eta = 0$ and $d\eta = 0$ by the positive definiteness of the scalar products $\langle\langle \cdot\,, \cdot \rangle\rangle$ on $\Omega^{k-1} M$ and $\Omega^{k+1} M$.

Notation. For an oriented closed Riemannian manifold M, let

$$\mathcal{H}^k M := \{\eta \in \Omega^k M : \delta\eta = 0 \text{ and } d\eta = 0\} = \{\eta \in \Omega^k M : \Delta\eta = 0\}$$

denote the vector space of harmonic k-forms on M.

Thus the first formula in our last corollary reads $\ker d_k = \operatorname{im} d_{k-1} \oplus \mathcal{H}^k M$, and since the kth de Rham cohomology of M was defined as $H^k M := \ker d_k / \operatorname{im} d_{k-1}$, we obtain the following as a corollary.

Hodge Theorem. *Every de Rham cohomology class of an oriented closed Riemannian manifold is represented by a well-defined harmonic form. More precisely: The canonical map*

$$\begin{aligned} \mathcal{H}^k &\longrightarrow H^k M, \\ \eta &\longmapsto [\eta] \end{aligned}$$

is an isomorphism for every k.

But it follows from

$$\begin{aligned} \ker d_k &= \operatorname{im} d_{k-1} \oplus \mathcal{H}^k M, \\ \Omega^k M &= \ker d_k \oplus \operatorname{im} \delta_{n-k-1} \end{aligned}$$

that

$$\Omega^k M = \operatorname{im} d_{k-1} \oplus \operatorname{im} \delta_{n-k-1} \oplus \mathcal{H}^k M.$$

This gives the following result.

Hodge Decomposition Theorem. *If M is an oriented closed Riemannian manifold, then*

$$\Omega^k M = d\Omega^{k-1} M \oplus \delta\Omega^{k+1} M \oplus \mathcal{H}^k M$$

as an orthogonal direct sum with respect to the scalar product given by

$$\langle\langle \eta, \zeta \rangle\rangle = \int_M \eta \wedge *\zeta.$$

12.6 Poincaré Duality

From the definition of the coderivative as "star-conjugate" to the Cartan derivative up to sign, it follows that $d * \eta = 0 \Leftrightarrow \delta\eta = 0$ and $\delta * \eta = 0 \Leftrightarrow d\eta = 0$. So the star operator gives an isomorphism

$$* : \mathcal{H}^k M \xrightarrow{\cong} \mathcal{H}^{n-k} M,$$

and hence, by the Hodge theorem, an isomorphism $H^k M \cong H^{n-k} M$. This is called Poincaré duality.

Theorem (Poincaré duality for de Rham cohomology). *If M is an oriented closed n-dimensional Riemannian manifold, then the star operator on harmonic forms defines an isomorphism $H^kM \cong H^{n-k}M$:*

$$\begin{array}{ccc} \mathcal{H}^kM & \xrightarrow{\cong} & H^kM \\ {\scriptstyle *}\downarrow{\scriptstyle \cong} & & {\scriptstyle \cong}\downarrow{\scriptstyle \text{Poincaré}} \\ \mathcal{H}^{n-k}M & \xrightarrow{\cong} & H^{n-k}M \end{array}$$

Incidentally, Poincaré duality says something interesting even for $k = 0$. For connected manifolds, as we recall (Section 11.2), $H^0M = \mathbb{R}$ canonically; so for orientable closed connected n-dimensional manifolds we also have $H^nM \cong \mathbb{R}$, and the choice of an orientation determines an isomorphism:

Corollary (of Poincaré duality). *If M is an oriented n-dimensional closed connected manifold, then the canonical homomorphism*

$$\begin{array}{ccc} H^nM & \longrightarrow & \mathbb{R}, \\ [\omega] & \longmapsto & \int_M \omega \end{array}$$

given by integration is an isomorphism.

Because of this, one might think that the nth de Rham cohomology for these manifolds would be as uninteresting as the zeroth. But one would be mistaken, because H^n, in contrast to H^0, acts nontrivially on *maps*.

Definition. If $f : M \to N$ is a differentiable map between oriented n-dimensional closed connected manifolds, then the well-defined number $\deg(f)$ given by

$$\int_M f^*\omega = \deg(f) \int_N \omega,$$

i.e. by the commutativity of

$$\begin{array}{ccc} H^nM & \xrightarrow[\cong]{\int_M} & \mathbb{R} \\ {\scriptstyle H^nf}\uparrow & & \uparrow{\scriptstyle \deg(f)} \\ H^nN & \xrightarrow[\int_N]{\cong} & \mathbb{R}, \end{array}$$

is called the ***mapping degree*** (or just the ***degree***) of f.

Of course, the degree of a constant map is zero (if $n > 0$). The same is true for any map that is not surjective, because then we can find an ω with $\int_N \omega \neq 0$ and $\operatorname{supp} \omega \subset N \setminus f(M)$, so $f^*\omega = 0$. The degree of an orientation-preserving (resp. orientation-reversing) diffeomorphism is $+1$ (resp. -1); see Section 5.5. The mapping degree is always an integer (Exercise 5.4), from which we might deduce its homotopy invariance if we didn't know it already because of the homotopy invariance of de Rham cohomology (Section 11.2).

We could also interpret the corollary above as a "converse of Stokes's theorem" for oriented closed manifolds: If $\int_M \omega = 0$, then $[\omega] = 0 \in H^nM$ and hence $\omega = d\alpha$.

12.7 Test

(1) Let V be a finite-dimensional real vector space, and let $\langle \cdot, \cdot \rangle$ denote a nondegenerate symmetric bilinear form given on V and also transferred canonically to V^*. Then, for $\varphi \in V^*$ and $v \in V$, we always have

☐ $\langle {}^\sharp\varphi, v \rangle + \langle \varphi, {}^\flat v \rangle = 0$.

☐ $\langle {}^\sharp\varphi, v \rangle = {}^\flat v({}^\sharp\varphi)$.

☐ $\langle \varphi, {}^\flat v \rangle = \varphi(v)$.

(2) Let V be a four-dimensional nondegenerate quadratic space of index 3. Then $\mathrm{Alt}^2 V$ is a six-dimensional nondegenerate quadratic space of index

☐ 3. ☐ 0. ☐ 6.

(3) Let $(M, \langle\cdot\,,\cdot\rangle)$ be an oriented semi-Riemannian manifold with volume form $\omega_M \in \Omega^n M$. How does multiplication of the metric by a positive function $\lambda : M \to \mathbb{R}^+$ affect the volume form?

- □ $\omega_{(M,\lambda\langle\cdot\,,\cdot\rangle)} = \omega_M$.
- □ $\omega_{(M,\lambda\langle\cdot\,,\cdot\rangle)} = \lambda^{\frac{n}{2}}\omega_M$.
- □ $\omega_{(M,\lambda\langle\cdot\,,\cdot\rangle)} = \lambda^{n}\omega_M$.

(4) Let V be a $2k$-dimensional nondegenerate quadratic space and assume that $k + \operatorname{index} M$ is *even*, so that the star operator on the forms of middle degree defines an *involution* $\mathrm{Alt}^k V \to \mathrm{Alt}^k V$; that is, $** = \mathrm{Id}$. Then the vector space $\mathrm{Alt}^k V$ is the direct sum of the subspaces of "self-dual" ($*\omega = \omega$) and "anti-self-dual" ($*\omega = -\omega$) alternating k-forms. Let s and a denote their dimensions. Are these dimensions independent of the index of the space V?

- □ Yes. The dimensions are always $a = s = \frac{1}{2}\binom{2k}{k}$.
- □ Yes, because the dimensions are just $a = 0$ or $s = 0$, depending on whether k is even or odd.
- □ No, because $s = 0$ in the negative definite case ($\operatorname{index} V = 2k$) but $a = 0$ in the positive definite case.

(5) What does the star operator do on an oriented semi-Riemannian manifold with the canonical volume form $\omega_M \in \Omega^n M$ and the constant 0-form $1 \in \Omega^0 M$?

- □ $*\omega_M = 1$ and $*1 = \omega_M$.
- □ $*\omega_M = (-1)^{\operatorname{index} M} 1$ and $*1 = \omega_M$.
- □ $*\omega_M = 1$ and $*1 = (-1)^{\operatorname{index} M}\omega_M$.

(6) The statement that $-\delta$ is formally adjoint to the Cartan derivative d means that

- □ $\int_M d\eta \wedge *\zeta + \int_M \eta \wedge *\delta\zeta = 0$
- □ $\int_M d\eta \wedge *\zeta + \int_M \eta \wedge \delta * \zeta = 0$

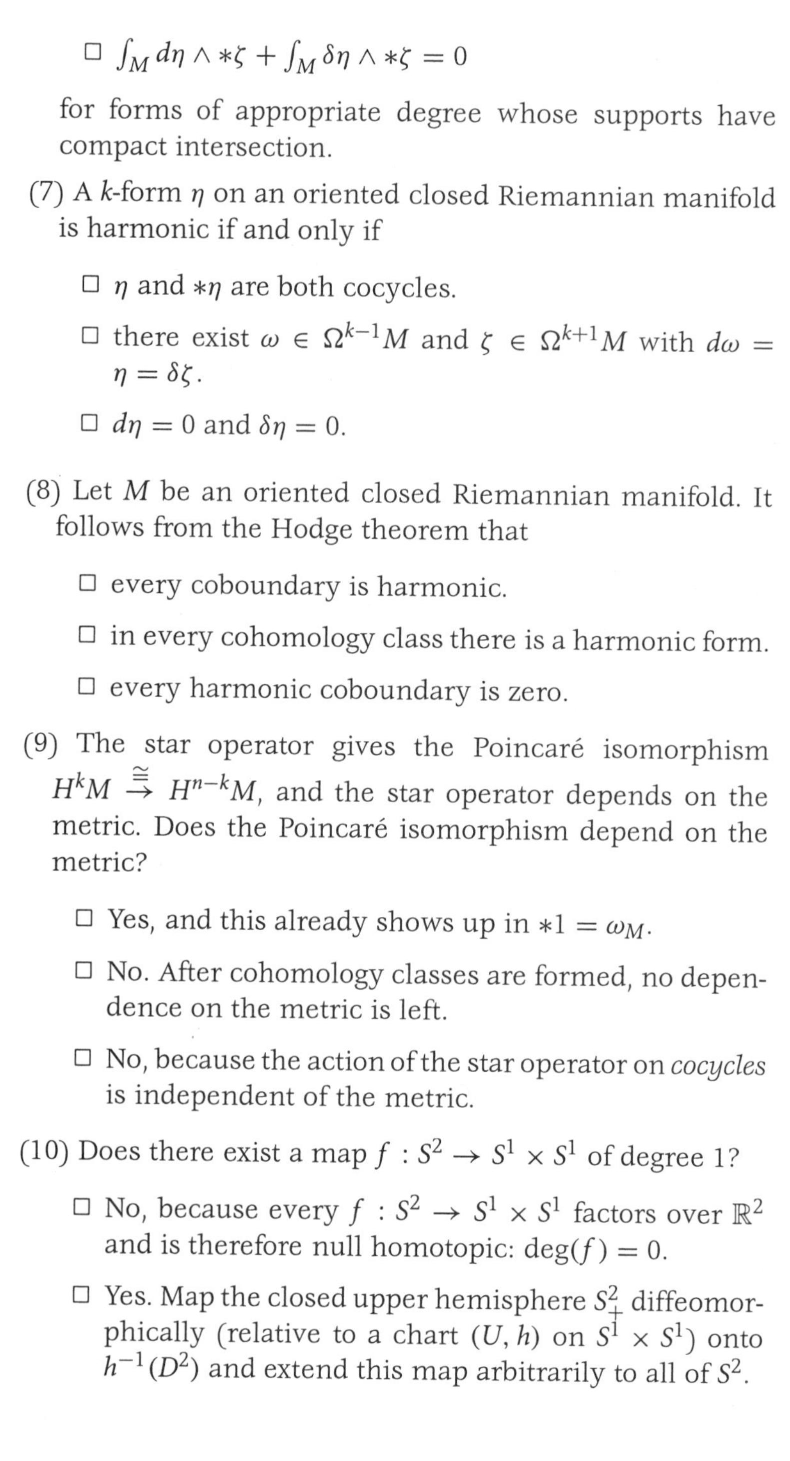

☐ $\int_M d\eta \wedge *\zeta + \int_M \delta\eta \wedge *\zeta = 0$

for forms of appropriate degree whose supports have compact intersection.

(7) A k-form η on an oriented closed Riemannian manifold is harmonic if and only if

☐ η and $*\eta$ are both cocycles.

☐ there exist $\omega \in \Omega^{k-1}M$ and $\zeta \in \Omega^{k+1}M$ with $d\omega = \eta = \delta\zeta$.

☐ $d\eta = 0$ and $\delta\eta = 0$.

(8) Let M be an oriented closed Riemannian manifold. It follows from the Hodge theorem that

☐ every coboundary is harmonic.

☐ in every cohomology class there is a harmonic form.

☐ every harmonic coboundary is zero.

(9) The star operator gives the Poincaré isomorphism $H^kM \overset{\cong}{\to} H^{n-k}M$, and the star operator depends on the metric. Does the Poincaré isomorphism depend on the metric?

☐ Yes, and this already shows up in $*1 = \omega_M$.

☐ No. After cohomology classes are formed, no dependence on the metric is left.

☐ No, because the action of the star operator on *cocycles* is independent of the metric.

(10) Does there exist a map $f : S^2 \to S^1 \times S^1$ of degree 1?

☐ No, because every $f : S^2 \to S^1 \times S^1$ factors over $\mathbb{R}^2$ and is therefore null homotopic: $\deg(f) = 0$.

☐ Yes. Map the closed upper hemisphere S^2_+ diffeomorphically (relative to a chart (U, h) on $S^1 \times S^1$) onto $h^{-1}(D^2)$ and extend this map arbitrarily to all of S^2.

□ Yes. In constructing such a map, take advantage of the fact that both the 2-sphere and the torus can arise from the square by identifying boundary points. Then the identity on the square induces a map of degree 1.

12.8 Exercises

Exercise 12.1. Let M be an oriented Riemannian manifold of dimension divisible by four, so that the star operator in the middle dimension is an *involution* $* : \Omega^{2k}M \to \Omega^{2k}M$; that is, $** = \mathrm{Id}$. In this case, a $2k$-form ω is called ***self-dual*** if $*\omega = \omega$ and ***anti-self-dual*** if $*\omega = -\omega$. Prove that every harmonic $2k$-form is, in a unique way, the sum of a self-dual and an anti-self-dual harmonic form.

Exercise 12.2. Again let M be an oriented Riemannian manifold, this time compact and without boundary. Consider the Laplace–de Rham operator $\Delta : \Omega^k M \to \Omega^k M$ on k-forms. We deviate from the sign convention of linear algebra in calling λ an ***eigenvalue*** of Δ if there exists a nonzero form $\omega \in \Omega^k M$ such that $\Delta\omega + \lambda\omega = 0$. To start with, this definition makes sense only for real λ, but since we can consider complex-valued k-forms $\omega + i\eta$ and apply Δ to real and imaginary parts, we may also ask about *complex* eigenvalues.

Prove that all the eigenvalues are real and greater than or equal to zero.

Exercise 12.3. Let $f : M \to N$ be a differentiable map between connected, oriented, compact n-dimensional manifolds without boundary, and let $q \in N$ be a regular value of f. Prove that

$$\deg(f) = \sum_{p \in f^{-1}(q)} \varepsilon(p),$$

where $\varepsilon(p) = \pm 1$ according to whether $df_p : T_pM \to T_{f(p)}N$ is orientation-preserving or orientation-reversing.

EXERCISE 12.4. Let $\pi : \widetilde{M} \to M$ be an r-sheeted cover (see, for instance, [J:*Top*], p. 130) of an n-dimensional manifold M. Then

$$(\pi_* \omega)_q := \sum_{p \in f^{-1}(q)} ((d\pi_p)^{-1})^* \omega_p$$

defines a homomorphism $\pi_* : \Omega^k \widetilde{M} \to \Omega^k M$, which induces a homomorphism $\pi_* : H^k \widetilde{M} \to H^k M$. Show that $\pi_* \circ \pi^* : H^k M \to H^k M$ is r times the identity, and conclude that the nth de Rham cohomology group vanishes for nonorientable compact connected n-dimensional manifolds.

12.9 Hints for the Exercises

FOR EXERCISE 12.1. Every $2k$-form is, in a unique way, the sum of a self-dual and an anti-self-dual form; this follows quite easily from $** = \mathrm{Id}$ and the linearity of the star operator, without looking further. But to show that for a harmonic form each of these two summands is itself harmonic, you have to take another look at the definition of the coderivative δ, which, along with learning what's going on, is the goal of this simple exercise.

FOR EXERCISE 12.2. As we saw in Section 12.5, $\langle\langle \cdot , \cdot \rangle\rangle$ turns $\Omega^k M$, under the given hypotheses, into a genuine Euclidean space, and in solving this problem you'll feel once again that you've been carried back to elementary linear algebra.

FOR EXERCISE 12.3. By the definition of the mapping degree in Section 12.6, it suffices to find *some* n-form that is tailored to the measure, has nonvanishing integral, and satisfies

$$\int_M f^* \omega = \sum_{p \in f^{-1}(q)} \varepsilon(p) \int_N \omega.$$

The support of such an ω will be set inside a sufficiently small neighborhood of q.

Every map has regular values, by the way; this follows from *Sard's theorem* (see, for instance, [BJ], Chapter 6). So when you prove the assertion of the exercise you'll prove simultaneously that the mapping degree defined by $H^n f$ is integer-valued. If you don't want to use Sard's theorem, you can stick to the homotopy invariance of $H^n f$, because it's quite easy to deform f homotopically in such a way that regular values occur.

FOR EXERCISE 12.4. The first part of the exercise (dealing with $\pi_* \circ \pi^*$) can be done straight from the definitions and would have fit well into Chapter 11. But Exercise 12.3 comes into play in the second part. Now you should consider the ***orientation double covering*** $\pi : \widetilde{M} \to M$, in which, as the name suggests, each $\pi^{-1}(x)$ consists of the two orientations of $T_x M$. This $\widetilde{M}$ is canonically oriented. Since M is compact, so is $\widetilde{M}$; since M is assumed to be nonorientable, $\widetilde{M}$ is connected. What is the mapping degree of the canonical sheet-interchanging involution $f : \widetilde{M} \to \widetilde{M}$, and what does it have to do with π^* and π_*?

13 CHAPTER Calculations in Coordinates

13.1 The Star Operator and the Coderivative in Three-Dimensional Euclidean Space

In this last chapter we examine how to calculate with the star operator and coderivative in local coordinates on semi-Riemannian manifolds. But first we pick up where we left off in Chapter 10 and consider the simple but important example $M = \mathbb{R}^3$ with the usual coordinates x^1, x^2, x^3, the usual orientation, and the usual scalar product (denoted by the multiplication symbol $\cdot$). The index is zero and $k(3-k)$ is always even, so by Note 3 in Section 12.3 the star operator is an involution: $** = \mathrm{Id}$. Note 1 in the same section gives the following:

Note. *For $M = \mathbb{R}^3$ as above, $*1 = dx^1 \wedge dx^2 \wedge dx^3 \in \Omega^3 M$. We also have $*dx^1 = dx^2 \wedge dx^3$ and its cyclic permutations*

$*dx^2 = dx^3 \wedge dx^1$ *and* $*dx^3 = dx^1 \wedge dx^2$. *In the notation of line, area, and volume elements as defined in Section 10.2, these equations become*

$$*1 = dV \quad (\text{and hence } *dV = 1),$$
$$*\vec{ds} = d\vec{S} \quad (\text{and hence } *d\vec{S} = \vec{ds}).$$

The sign of the coderivative $\delta = \pm * d*^{-1}$ given in its definition in Section 12.4 is exactly the one that makes

$$\begin{array}{ccccccccc}
0 & \longrightarrow & \Omega^0 M & \xrightarrow{d} & \Omega^1 M & \xrightarrow{d} & \Omega^2 M & \xrightarrow{d} & \Omega^3 M & \longrightarrow & 0 \\
 & & *\downarrow\cong & & *\downarrow\cong & & *\downarrow\cong & & *\downarrow\cong & & \\
0 & \longrightarrow & \Omega^3 M & \xrightarrow[\delta]{} & \Omega^2 M & \xrightarrow[-\delta]{} & \Omega^1 M & \xrightarrow[\delta]{} & \Omega^0 M & \longrightarrow & 0
\end{array}$$

commutative. The translation of the Cartan derivative into grad, div, and curl (see Section 10.3) gives us another formulation of the commutativity of the diagram.

Note.

$$\delta(\vec{a} \cdot \vec{ds}) = * d(\vec{a} \cdot d\vec{S}) = * \operatorname{div} \vec{a}\, dV = \operatorname{div} \vec{a},$$
$$\delta(\vec{b} \cdot d\vec{S}) = - * d(\vec{b} \cdot \vec{ds}) = - * \operatorname{curl} \vec{b} \cdot d\vec{S} = -\operatorname{curl} \vec{b} \cdot \vec{ds},$$
$$\delta(c\, dV) = * dc = * \operatorname{grad} c \cdot \vec{ds} = \operatorname{grad} c \cdot d\vec{S}.$$

Note. *In terms of the translation isomorphisms of Section 10.2, this says that for any open subset X of $\mathbb{R}^3$, the following diagram is commutative:*

$$\begin{array}{ccccccccc}
0 & \longrightarrow & \Omega^3 X & \xrightarrow{\delta} & \Omega^2 X & \xrightarrow{\delta} & \Omega^1 X & \xrightarrow{\delta} & \Omega^0 X & \longrightarrow & 0 \\
 & & \downarrow\cong & & \downarrow\cong & & \downarrow\cong & & \downarrow\cong & & \\
0 & \longrightarrow & C^\infty(X) & \xrightarrow[\text{grad}]{} & \mathcal{V}(X) & \xrightarrow[-\text{curl}]{} & \mathcal{V}(X) & \xrightarrow[\text{div}]{} & C^\infty(X) & \longrightarrow & 0
\end{array}$$

In the classical notation and written on one line, the sequences for the Cartan derivative (above) and coderivative read as follows:

$$0 \rightleftarrows C^\infty(X) \underset{\text{div}}{\overset{\text{grad}}{\rightleftarrows}} \mathcal{V}(X) \underset{-\text{curl}}{\overset{\text{curl}}{\rightleftarrows}} \mathcal{V}(X) \underset{\text{grad}}{\overset{\text{div}}{\rightleftarrows}} C^\infty(X) \rightleftarrows 0$$

Setting $\Delta_X := d\delta + \delta d$ gives our next result.

Corollary. *For an open subset X of $\mathbb{R}^3$, the Laplace–de Rham operator in the classical notation is given as follows:*

(i) *For 0-forms and 3-forms,*

$$\Delta_X = \operatorname{div}\operatorname{grad} : C^\infty(X) \longrightarrow C^\infty(X).$$

(ii) *For 1-forms and 2-forms,*

$$\Delta_X = \operatorname{grad}\operatorname{div} - \operatorname{curl}\operatorname{curl} : \mathcal{V}(X) \longrightarrow \mathcal{V}(X).$$

Note that

$$\operatorname{div}\operatorname{grad} = \sum_{i=1}^{3} \frac{\partial^2}{\partial x^{i2}} = \Delta$$

is the classical Laplacian. So the sign convention for the coderivative that we fixed in the definition in Section 12.4, the one that makes δ the formal adjoint of $-d$, is consistent—at least in this respect—with the usual notation.

13.2 Forms and Dual Forms on Manifolds without a Metric

The language of computation in coordinates is the Ricci calculus, which will be discussed in detail in the present chapter. We last dealt with the Ricci calculus in Section 2.8, and encountered some of its principles there in the example of tangent vectors and vector fields. Vector fields and 1-forms are in a certain sense dual to each other, and we have meanwhile generalized the 1-forms or Pfaffian forms to k-forms. For a systematic description of the Ricci calculus in the framework of the Cartan calculus, it is convenient to generalize vector fields to "dual k-forms" in a similar way. This has nothing to do yet with orientation and metric, so we just consider an n-dimensional manifold M and a chart (U, h) on it.

Vector fields, 1-forms, and k-forms on U can then be written uniquely as

$$\begin{aligned} v &= \sum_{\mu=1}^{n} v^{\mu} \partial_{\mu}, \\ \omega &= \sum_{\mu=1}^{n} \omega_{\mu} dx^{\mu}, \\ \omega &= \sum_{\mu_1 < \cdots < \mu_k} \omega_{\mu_1 \ldots \mu_k} dx^{\mu_1} \wedge \ldots \wedge dx^{\mu_k}, \end{aligned}$$

respectively, where the components v^{μ}, ω_{μ}, and $\omega_{\mu_1 \ldots \mu_k}$ are real functions on U.

A k-form ω on M assigns to every $p \in M$ an alternating k-form $\omega_p \in \mathrm{Alt}^k T_p M$. For the definition of dual k-forms, $T_p M$ is simply replaced by the dual space $T_p^* M$.

Definition. A ***dual k-form*** on a manifold M is an assignment to every $p \in M$ of an alternating k-form $w_p \in \mathrm{Alt}^k T_p^* M$ on the dual space $T_p^* M = \mathrm{Hom}(T_p M, \mathbb{R})$ of the tangent space. The vector space of differentiable (relative to charts) dual k-forms on M will be denoted by $\Omega_k M$.

For finite-dimensional vector spaces V, we have $V^{**} = V$ canonically, so $\mathrm{Alt}^1 T_p^* M = T_p M$. Thus the dual 1-forms are the same as vector fields, and by analogy with $v = \sum_{\mu=0}^{n} v^{\mu} \partial_{\mu}$, we write the dual k-forms in local coordinates as follows.

Note and Notation. *If (U, h) is a chart, then every dual k-form w on U can be written uniquely as*

$$w = \sum_{\mu_1 < \cdots < \mu_k} w^{\mu_1 \ldots \mu_k} \partial_{\mu_1} \wedge \ldots \wedge \partial_{\mu_k}.$$

13.3 Three Principles of the Ricci Calculus on Manifolds without a Metric

We now use these objects—k-forms and dual k-forms, and 1-forms and vector fields in particular—to illustrate three general principles of the Ricci calculus:

(1) the description of objects by their components;

(2) the position of indices according to behavior under transformations;

(3) the summation convention.

For (1). Recall that what is meant by a contravariant vector v^μ in the Ricci calculus is the vector field $v = \sum v^\mu \partial_\mu$. Similarly, a ***covariant vector*** a_μ is to be understood as the 1-form $\sum a_\mu dx^\mu$. We extend these notions further by defining a ***skew-symmetric***, or ***alternating***, ***covariant tensor*** $\omega_{\mu_1 \dots \mu_k}$ of degree (or rank) k as the k-form

$$\sum_{\mu_1 < \dots < \mu_k} \omega_{\mu_1 \dots \mu_n} dx^{\mu_1} \wedge \dots \wedge dx^{\mu_k},$$

and an ***alternating contravariant tensor*** $w^{\mu_1 \dots \mu_k}$ of degree (or rank) k as the dual k-form

$$\sum_{\mu_1 < \dots < \mu_k} w^{\mu_1 \dots \mu_n} \partial_{\mu_1} \wedge \dots \wedge \partial_{\mu_k}.$$

So, for example, if you encounter a covariant skew-symmetric field tensor $F_{\mu\nu}$ of rank two in the physics literature, then as a mathematical reader you should realize that this means the 2-form $\sum_{\mu<\nu} F_{\mu\nu} dx^\mu \wedge dx^\nu$, because you won't be reminded of it.

Finally, we mustn't forget that in the notation and language of the Ricci calculus, no distinction is made between a geometric object ω and its restriction $\omega|U$ to a chart domain, so the component symbol takes on the additional job of denoting the entire object whenever necessary.

For (2). Whether the coordinate index of a component function is written as a superscript or a subscript is not left to chance in the Ricci calculus, but is determined by the behavior of the components under a change of charts. As in Exercise 3.3 (see the hint for this exercise on p. 63), let's denote the new coordinates by $x^{\overline{1}}, \ldots, x^{\overline{n}}$—defined without loss of generality on the same chart domain (suppressed in the notation anyway) as $x^1, \ldots, x^n$—and, savoring the double meaning of the $x^1, \ldots, x^n$ as functions on $U \subset M$ and as coordinates in $\mathbb{R}^n$, let's write the transition map as

$$\begin{aligned} x^{\overline{1}} &= x^{\overline{1}}(x^1, \ldots, x^n) \\ &\vdots \\ x^{\overline{n}} &= x^{\overline{n}}(x^1, \ldots, x^n) \end{aligned}$$

and its Jacobian matrix as

$$\left(\frac{\partial x^{\overline{\mu}}}{\partial x^{\mu}} \right)_{\overline{\mu}, \mu = 1, \ldots, n}.$$

Always keeping in mind that an x with a barred index means something completely different from an x with an unbarred index, we see from the chain rule that the following relations hold.

Note. *Under a change of coordinates,*

$$dx^{\overline{\mu}} = \sum_{\mu=1}^{n} \frac{\partial x^{\overline{\mu}}}{\partial x^{\mu}} dx^{\mu} \quad \textit{and} \quad \partial_{\overline{\mu}} = \sum_{\mu=1}^{n} \frac{\partial x^{\mu}}{\partial x^{\overline{\mu}}} \partial_{\mu}.$$

Corollary. *Under a change of coordinates, the component functions of k-forms and dual k-forms transform as follows:*

$$\omega_{\overline{\mu}_1 \ldots \overline{\mu}_k} = \sum_{\mu_1, \ldots, \mu_k = 1}^{n} \frac{\partial x^{\mu_1}}{\partial x^{\overline{\mu}_1}} \cdot \ldots \cdot \frac{\partial x^{\mu_k}}{\partial x^{\overline{\mu}_k}} \omega_{\mu_1 \ldots \mu_k}$$

$$w^{\overline{\mu}_1 \ldots \overline{\mu}_k} = \sum_{\mu_1, \ldots, \mu_k = 1}^{n} \frac{\partial x^{\overline{\mu}_1}}{\partial x^{\mu_1}} \cdot \ldots \cdot \frac{\partial x^{\overline{\mu}_k}}{\partial x^{\mu_k}} w^{\mu_1 \ldots \mu_k}.$$

Because we are actually summing over all multi-indices $(\mu_1, \ldots, \mu_k)$ here, not just the ordered ones, let me remind you that the component functions $\omega_{\mu_1\ldots\mu_k}$ and $w^{\mu_1\ldots\mu_k}$ are defined for all multi-indices, even though (because of the alternating property) the components with ordered indices $\mu_1 < \cdots < \mu_k$ already contain all the information.

Thus in each transformation formula the summation indices on the right-hand side are in opposite positions, and the free indices are in the same position on the right-hand side as on the left-hand side.

For (3). In the Ricci calculus, sums in which the index of summation appears twice, once as a subscript and once as a superscript (with the understanding that a superscript in the denominator acts as a subscript) occur often enough to have led to the adoption of the ***Einstein summation convention***, according to which one still thinks of the summation symbol but no longer writes it down. When the summation convention is applied, terms such as

$$v^\mu \partial_\mu, \quad a_\mu dx^\mu, \quad \text{and} \quad \frac{\partial x^\mu}{\partial x^{\overline{\mu}}} \frac{\partial x^\nu}{\partial x^{\overline{\nu}}} A_{\mu\nu}$$

are automatically read as

$$\sum_{\mu=1}^{n} v^\mu \partial_\mu, \quad \sum_{\mu=1}^{n} a_\mu dx^\mu, \quad \text{and} \quad \sum_{\nu=1}^{n} \sum_{\mu=1}^{n} \frac{\partial x^\mu}{\partial x^{\overline{\mu}}} \frac{\partial x^\nu}{\partial x^{\overline{\nu}}} A_{\mu\nu}$$

if nothing is explicitly said otherwise. For example, we are explicitly told to deal differently with

$$\sum_{\mu_1 < \cdots < \mu_k} \omega_{\mu_1\ldots\mu_k} dx^{\mu_1} \wedge \ldots \wedge dx^{\mu_k}.$$

But if we wanted to, we could also use the summation convention to write this representation of a k-form in local coordinates without the summation sign, namely as

$$\sum_{\mu_1 < \cdots < \mu_k} \omega_{\mu_1\ldots\mu_k} dx^{\mu_1} \wedge \ldots \wedge dx^{\mu_k} = \frac{1}{k!} \underbrace{\omega_{\mu_1\ldots\mu_k} dx^{\mu_1} \wedge \ldots \wedge dx^{\mu_k}}_{\text{summation convention}}.$$

13.4 Tensor Fields

The discussion in the preceding section also applies to the still more general ***tensors of covariant degree r and contravariant degree s*** of the Ricci calculus, which have not yet been introduced. The component functions of these tensors have r subscripts and s superscripts, and the ordering of these indices also matters if no symmetry requirements are imposed. As an example, consider the type of tensor in which the r subscripts come first. The component functions—and in the Ricci calculus the whole tensor as well—are then written, for instance, as

$$A_{\mu_1 \ldots \mu_r}{}^{\nu_1 \ldots \nu_s}.$$

From this position of the indices, the expert in the Ricci calculus concludes that under a change of coordinates the tensor transforms as

$$A_{\overline{\mu}_1 \ldots \overline{\mu}_r}{}^{\overline{\nu}_1 \ldots \overline{\nu}_r} = \sum_{\substack{\text{all } \nu \\ \text{all } \mu}} \frac{\partial x^{\mu_1}}{\partial x^{\overline{\mu}_1}} \cdot \ldots \cdot \frac{\partial x^{\mu_r}}{\partial x^{\overline{\mu}_r}} \cdot \frac{\partial x^{\overline{\nu}_1}}{\partial x^{\nu_1}} \cdot \ldots \cdot \frac{\partial x^{\overline{\nu}_s}}{\partial x^{\nu_s}} A_{\mu_1 \ldots \mu_r}{}^{\nu_1 \ldots \nu_s}.$$

This simultaneously defines what such a tensor of covariant degree r and contravariant degree s "is" in the Ricci calculus, in case anyone should ask. In fact, we can safely settle for this definition; we already saw in Chapter 1 how to make it precise for $r = 0$ and $s = 1$ ("physically defined" tangent vectors and vector fields).

Whoever is still unsatisfied can get a conceptually better answer from multilinear algebra to the question "What is a tensor?" The coordinate-independent object A of which the $A_{\mu_1 \ldots \mu_r}{}^{\nu_1 \ldots \nu_s}$ are only the component functions is

$$A = \sum_{\substack{\text{all } \nu \\ \text{all } \mu}} A_{\mu_1 \ldots \mu_r}{}^{\nu_1 \ldots \nu_s} dx^{\mu_1} \otimes \cdots \otimes dx^{\mu_r} \otimes \partial_{\nu_1} \otimes \cdots \otimes \partial_{\nu_s},$$

which assigns to every p an element

$$A(p) \in T_p^*M \otimes \cdots \otimes T_p^*M \otimes T_pM \otimes \cdots \otimes T_pM.$$

If the ordering of the r superscripts and s subscripts is changed, then the ordering of the factors in the tensor product changes accordingly.

But what does the mysterious symbol $\otimes$ mean? It would be nice if all mathematics students learned this in second-semester linear algebra. Well, you'll learn it someday, and then you'll see the tensors of the Ricci calculus with different eyes ....

But I won't make it quite so easy for myself. I'm going to give you a short minicourse, a *microcourse*, on the tensor product. Pay attention, it's starting: The first thing you have to know about the tensor product of two (and similarly of several) vector spaces V and W is that it's really a *pair* $(V \otimes W, t)$ consisting of a vector space $V \otimes W$ and an operation $t : V \times W \to V \otimes W$, with the operation denoted by $(v, w) \mapsto v \otimes w$. Thus one can also take tensor products of individual vectors, and these tensor products are elements of the tensor product of the spaces. Be careful, though: in general, the tensor product of the spaces is *not* the set of tensor products of their elements; the operation is not surjective. So it's not as you might think, that if you just understand $v \otimes w$ you'll automatically know $V \otimes W$. In fact you can't actually understand either $v \otimes w$ or $V \otimes W$ by itself; you really have to look at the pair $(V \otimes W, t)$.

So after all that can we just write down the map $t : V \times W \to V \otimes W$? We could. The question is whether you'd get much out of it. At the moment, I'd rather start by telling you something more important: The operation t is *universally bilinear* in the sense that first, of course it's bilinear itself, as any decent product ought to be, and second, every bilinear map on $V \times W$ arises in exactly one way from post-composition with a linear map on $V \otimes W$; more precisely, for every bilinear map $f : V \times W \to X$ there is exactly one linear map $\varphi : V \otimes W \to X$ such that $f = \varphi \circ t$.

Admittedly, whether there *exists* a pair $(V \otimes W, t)$ with this marvelous universal property is something I haven't proved yet, but you can already see that there can be essentially *at most* one. For if $(V\widetilde{\otimes}W, \widetilde{t})$ is another pair, we can play the universal property of $\widetilde{t}$ off against t, and vice versa, to obtain linear maps $V \otimes W \rightleftarrows V\widetilde{\otimes}W$ that are compatible with t and $\widetilde{t}$ and are inverses of each other. This is also the reason it isn't so important *how* a universal $(V \otimes W, t)$ is constructed as long as such a construction is possible at all.

And it *is* possible. Here's how to see it: any arbitrary set A generates the real vector space $F(A)$ of *formal linear combinations* $c_1a_1 + \cdots + c_ka_k$, whose elements are actually the maps $c : A \to \mathbb{R}$ that send all but finitely many $a \in A$ to zero, but which are written, for practical reasons, as sums $\sum c(a)a$ as above. Through $a \mapsto 1a$ we also have a canonical map $A \to F(A)$, and this map $V \times W \to F(V \times W)$ is precisely the one we now consider for the special case $A := V \times W$. It has a universal property, but not the right one yet; it isn't even bilinear. So now we touch it up in a completely routine way. That is, we consider all the elements in $F(V \times W)$ that have one of the two forms

(a) $\quad (c_1v_1 + c_2v_2, w) - c_1(v_1, w) - c_2(v_2, w)$

(b) $\quad (v, c_1w_1 + c_2w_2) - c_1(v, w_1) - c_2(v, w_2)$

—these are the elements whose nonvanishing is an obstruction to bilinearity—and take the quotient of $F(V \times W)$ by the subspace $F_0 \subset F(V \times W)$ they generate. Then the quotient $V \otimes W := F(V \times W)/F_0$ and the canonical map

$$t : V \times W \to F(V \times W) \to F(V \times W)/F_0$$

together form a universal bilinear pair for V and W, as desired.

But you need this construction only if you're stopped by the police and have to justify your use of the tensor product. For your daily work, you're better off deriving what you want to know about the tensor product directly from the universal property.

End of the microcourse! You'll have to admit that it was quick enough to read through. Granted, you don't have a firm seat in the tensor saddle yet. That would take a whole swarm of trivial but not superfluous lemmas, for which unfortunately there's no room in my book.

Since the alternating forms are multilinear, of course they have something to do with the tensor product. I'll mention only that canonically

$$\text{Alt}^k T_pM \subset \underbrace{T_p^*M \otimes \cdots \otimes T_p^*M}_{k},$$

so every k-form ω is also a covariant tensor of degree k in this general sense, where, reassuringly, the component functions are the same in both interpretations. From the skew symmetry in the indices, it actually follows that

$$\sum_{\mu_1 < \cdots < \mu_k} \omega_{\mu_1 \ldots \mu_k} dx^{\mu_1} \wedge \ldots \wedge dx^{\mu_k} = \sum_{\text{all } \mu} \omega_{\mu_1 \ldots \mu_k} dx^{\mu_1} \otimes \cdots \otimes dx^{\mu_k}$$

according to our normalization of the wedge product (see the theorem in Section 8.2). So in passing to the more general concept of tensors in the Ricci calculus, you don't have to learn any new conventions for the old familiar k-forms.

13.5 Raising and Lowering Indices in the Ricci Calculus

These three notational conventions of the Ricci calculus—(1) description by components, (2) position of the indices, and (3) the summation convention—refer to calculating in coordinates on an n-dimensional manifold M without additional structure. But if a semi-Riemannian metric $\langle \cdot , \cdot \rangle$ is given on M, a fourth convention is added—one that deals with the celebrated "raising and lowering" of indices. We begin by considering the procedure completely formally, and only afterwards do we ask about its mathematical content.

Notation (Raising and lowering indices in the Ricci calculus). Let $(M, \langle\cdot,\cdot\rangle)$ be a semi-Riemannian manifold. In local coordinates, we write $g_{\mu\nu} := \langle\partial_\mu, \partial_\nu\rangle$ as usual and let $(g^{\mu\nu})$ denote the inverse of the matrix $(g_{\mu\nu})$. Now let A be a tensor of covariant degree r and contravariant degree s, written in the Ricci calculus with $r + s$ indices. Without loss of generality, let one index be ν and none be μ. Then, according to whether ν is a subscript or a superscript, we write

$$A\cdots{}^{\mu}\cdots := g^{\mu\nu}A\cdots{}_{\nu}\cdots \quad \text{or} \quad A\cdots{}_{\mu}\cdots := g_{\mu\nu}A\cdots{}^{\nu}\cdots,$$

where the summation convention is to be applied. There is no change in either the position of or the notation for the remaining indices, of whose presence you should be reminded by the dots.

So if a contravariant vector v^μ is given, for instance, the notation v_μ is no longer free. By this convention, it now means $g_{\mu\nu}v^\nu$. Here are some more examples, just to get you used to the formal procedure:

$$\begin{aligned} A^\mu &= g^{\mu\nu}A_\nu, \\ F_\mu^\nu &= g^{\nu\lambda}F_{\mu\lambda} = g_{\mu\lambda}F^{\lambda\nu}, \\ F^{\mu\nu} &= g^{\mu\lambda}g^{\nu\kappa}F_{\lambda\kappa}, \\ \omega^{\mu_1\ldots\mu_k} &= g^{\mu_1\nu_1}\cdot\ldots\cdot g^{\mu_k\nu_k}\omega_{\nu_1\ldots\nu_k}. \end{aligned}$$

Of course, you may already guess that the result of raising and lowering indices is another tensor—that, under a change of coordinates, the newly created indexed quantities transform correctly according to the (new) position of the indices. Otherwise the Ricci calculus would hardly have settled on this convention. To check this, observe first that $g_{\mu\nu} = \langle\partial_\mu, \partial_\nu\rangle$ transforms correctly as a covariant tensor of degree two, by the note in Section 13.3. For every $p \in M$, $\langle\cdot,\cdot\rangle_p$ is a bilinear form on T_pM and thus an element of $(T_pM \otimes T_pM)^* = T_p^*M \otimes T_p^*M$; the $g_{\mu\nu}$ are the component functions of this "fundamental tensor" (as it is called in the Ricci calculus) of the semi-Riemannian manifold. Hence $g^{\mu\nu}$

also transforms as a contravariant tensor. The assertion follows by direct substitution and calculation, using the fact that the Jacobian matrices of the two transition maps (from the old coordinates to the new and back) are inverses of each other in a natural way.

The processes of raising and lowering a given index are themselves inverses of each other, since

$$g_{\lambda\mu}g^{\mu\nu} = g^{\lambda\mu}g_{\mu\nu} = \begin{cases} 1 & \text{for } \lambda = \nu, \\ 0 & \text{otherwise} \end{cases}$$

by definition. Because the matrices are symmetric, it also follows that

$$g^{\mu\nu} = g^{\mu\lambda}g^{\nu\kappa}g_{\lambda\kappa},$$

so the notation $(g^{\mu\nu})$ for the inverse matrix of $(g_{\mu\nu})$ is consistent with the convention: raising the two indices really does turn $g_{\mu\nu}$ into $g^{\mu\nu}$.

In general, indices should not be stacked on top of each other because we want the overall order of the indices to stay recognizable. But as long as no indices are raised or lowered, no misunderstandings occur within the Ricci calculus if the separate orderings of the upper and lower indices are known. If $A_{\mu\nu}$ is *symmetric* in μ and ν, for instance, then of course $A_\mu{}^\nu = A^\nu{}_\mu$ for the component functions, so in computations we just write A_μ^ν.

13.6 The Invariant Meaning of Raising and Lowering Indices

Now, how can raising and lowering indices be understood conceptually and in a coordinate-free way? To answer this question, we consider for each $p \in M$ the isomorphism

$$\begin{aligned} T_pM &\overset{\cong}{\longrightarrow} T_p^*M, \\ v &\longmapsto \langle v, \cdot \rangle \end{aligned}$$

between the tangent and cotangent spaces, which is determined by the semi-Riemannian metric and for which we introduced the notation

$$T_pM \underset{\sharp}{\overset{\flat}{\rightleftarrows}} T_p^*M$$

in Section 12.2. How does this look in local coordinates? As we know, for any 1-form $\omega = \omega_\mu dx^\mu$ the νth component function is given by $\omega_\nu = \omega(\partial_\nu)$. This gives us an answer in the special case $\omega = {}^\flat\partial_\mu := \langle \partial_\mu, \cdot \rangle$:

Note. *We have*

$${}^\flat\partial_\mu = \langle \partial_\mu, \partial_\nu \rangle dx^\nu = g_{\mu\nu} dx^\nu,$$

and hence also

$${}^\sharp dx^\mu = g^{\mu\nu}\partial_\nu.$$

Corollary. *Converting contravariant to covariant vectors and vice versa by lowering and raising indices, respectively, in the Ricci calculus corresponds to the isomorphism* $\flat : T_pM \to T_p^*M$ *and its inverse* $\sharp$ *given by the semi-Riemannian metric. More precisely,*

$$\begin{aligned} {}^\flat(v^\mu \partial_\mu) &= v_\mu dx^\mu, \\ {}^\sharp(a_\mu dx^\mu) &= a^\mu \partial_\mu. \end{aligned}$$

Similarly, we have the more general observation: Applying $\flat$ or $\sharp$ to the ith factor of a tensor product of degree $(r+s)$, with r factors in T_p^*M and s factors in T_pM (in a specific ordering), is described in the Ricci calculus by lowering or raising, respectively, the ith of the $r+s$ tensor indices. For example, under

$$\begin{array}{c} T_pM \otimes T_p^*M \otimes T_pM \\ \cong \Big\downarrow \flat \otimes \mathrm{Id} \otimes \mathrm{Id} \\ T_p^*M \otimes T_p^*M \otimes T_pM \end{array}$$

the tensor $A^\lambda{}_\mu{}^\nu$ of covariant degree one and contravariant degree two goes to the tensor $A_{\lambda\mu}{}^\nu$ of covariant degree two and contravariant degree one (in the sense of convention (1) of the Ricci calculus, of course—it would be meaningless to assign, say, the component $A_{11}{}^{2}$ to the individual component $A^{1}{}_{1}{}^{2}$). This is true because $\flat \otimes \mathrm{Id} \otimes \mathrm{Id}$ sends $A^{\lambda}{}_{\mu}{}^{\nu} \partial_\lambda \otimes dx^\mu \otimes \partial_\nu$ to $A^{\lambda}{}_{\mu}{}^{\nu} ({}^{\flat}\partial_\lambda) \otimes dx^\mu \otimes \partial_\nu$, which is

$$A^{\lambda}{}_{\mu}{}^{\nu} g_{\lambda\sigma} dx^\sigma \otimes dx^\mu \otimes \partial_\nu = A_{\sigma\mu}{}^{\nu} dx^\sigma \otimes dx^\mu \otimes \partial_\nu$$

by the corollary above.

Raising all the indices of a k-form produces a dual k-form, and vice versa. These procedures are also given in a coordinate-free way through $\sharp$ and $\flat$ as

$$\begin{array}{c} \mathrm{Alt}^k T_p M \\ {\scriptstyle \mathrm{Alt}^k \flat} \big\uparrow \big\downarrow {\scriptstyle \mathrm{Alt}^k \sharp} \\ \mathrm{Alt}^k T_p^* M. \end{array}$$

In other words, the diagram

$$\begin{array}{ccc} \mathrm{Alt}^k T_p M & \longrightarrow & T_p^* M \otimes \cdots \otimes T_p^* M \\ \big\downarrow {\scriptstyle \mathrm{Alt}^k \sharp} & & \big\downarrow {\scriptstyle \sharp \otimes \cdots \otimes \sharp} \\ \mathrm{Alt}^k T_p^* M & \longrightarrow & T_p M \otimes \cdots \otimes T_p M \end{array}$$

commutes.

13.7 Scalar Products of Tensors in the Ricci Calculus

The notation for raising and lowering indices is very convenient for computations with the various scalar products we have to consider. Of course, for tangent vectors themselves, the following holds because $\langle \partial_\mu, \partial_\nu \rangle =: g_{\mu\nu}$.

Note. *For vector fields v and w, we have*

$$\langle v, w \rangle = g_{\mu\nu} v^\mu w^\nu = v_\mu w^\mu$$

in local coordinates.

According to the definition, the isomorphism $\flat$ transfers the scalar product of T_pM to T_p^*M (special case of the defining lemma for the scalar product in the space of forms; see Section 12.2). This gives the following.

Note. *For 1-forms α and β, we have*

$$\langle \alpha, \beta \rangle = \alpha_\mu \beta^\mu = g^{\mu\nu} \alpha_\mu \beta_\nu$$

in local coordinates. In particular, $\langle dx^\mu, dx^\nu \rangle = g^{\mu\nu}$.

Although we called the scalar product on $\mathrm{Alt}^k T_pM$ *canonical* in the definition, we don't want to forget that the wedge product was part of that definition, and the normalization of the wedge product in the literature is not completely uniform. For this reason, we always have to be careful about checking even plausible scalar product formulas for k-forms.

Lemma. *For k-forms $\eta, \zeta \in \Omega^k M$ on a semi-Riemannian manifold, we have*

$$\langle \eta, \zeta \rangle = \sum_{\mu_1 < \cdots < \mu_k} \eta_{\mu_1 \ldots \mu_k} \zeta^{\mu_1 \ldots \mu_k} = \frac{1}{k!} \eta_{\mu_1 \ldots \mu_k} \zeta^{\mu_1 \ldots \mu_k}$$

in local coordinates.

PROOF. First, the definition of the scalar product in Section 12.2 implies that

$$\langle \eta, {}^\flat\partial_{\mu_1} \wedge \ldots \wedge {}^\flat\partial_{\mu_k} \rangle = \eta(\partial_{\mu_1}, \ldots, \partial_{\mu_k}) =: \eta_{\mu_1 \ldots \mu_k}.$$

But lowering indices, as we explained earlier, has the same effect as applying $\flat$. So we can write ζ as

$$\zeta = \sum_{\mu_1 < \cdots < \mu_k} \zeta^{\mu_1 \ldots \mu_k} \, {}^\flat\partial_{\mu_1} \wedge \ldots \wedge {}^\flat\partial_{\mu_k},$$

and the assertion follows. □

With a clear conscience, we can describe the scalar product on the tensor product $V \otimes W$ of two quadratic spaces $(V, \langle \cdot, \cdot \rangle_V)$ and $(W, \langle \cdot, \cdot \rangle_W)$ as *given canonically*. It is the bilinear form on $V \otimes W$ that satisfies

$$\langle v \otimes w, v' \otimes w' \rangle = \langle v, v' \rangle_V \langle w, w' \rangle_W,$$

and similarly for tensor products with several factors. In particular, at every point p of a semi-Riemannian manifold M, there is a scalar product for tensors of covariant degree r and contravariant degree s. For the type of tensor in which all r covariant factors appear first, for example, we have

$$\langle A, B \rangle = A_{\mu_1 \dots \mu_r}{}^{\nu_1 \dots \nu_s} B^{\mu_1 \dots \mu_r}{}_{\nu_1 \dots \nu_s}$$

in local coordinates. Thus interpreting k-forms as covariant tensors of degree k through $\mathrm{Alt}^k T_p M \subset T_p^* M \otimes \cdots \otimes T_p^* M$ leads to another scalar product:

$$\langle \eta, \zeta \rangle_{\text{scalar product of } k\text{-forms}} = \frac{1}{k!} \langle \eta, \zeta \rangle_{\text{scalar product of tensors}}.$$

Well, you can't have everything! We'll continue to use the scalar product on k-forms anyway.

13.8 The Wedge Product and the Star Operator in the Ricci Calculus

Now let M be an oriented n-dimensional semi-Riemannian manifold. How do the star operator and coderivative look in the Ricci calculus? Since $\eta \wedge *\zeta = \langle \eta, \zeta \rangle \omega_M$, we begin by examining the wedge product and the volume form. For $\omega \in \Omega^r M$ and $\eta \in \Omega^s M$, we have

$$\omega \wedge \eta = \sum_{\substack{\mu_1 < \cdots < \mu_r \\ \nu_1 < \cdots < \nu_s}} \omega_{\mu_1 \dots \mu_r} \eta_{\nu_1 \dots \nu_s} dx^{\mu_1} \wedge \ldots \wedge dx^{\mu_r} \wedge dx^{\nu_1} \wedge \ldots \wedge dx^{\nu_s}$$

in local coordinates. From this we can read off a formula for the components $(\omega \wedge \eta)_{\mu_1 \dots \mu_{r+s}}$, $\mu_1 < \cdots < \mu_{r+s}$, of the wedge product. In order to write it down we interpret a partition of the set $\{1, \dots, r+s\}$ into one subset of r elements and another of s elements as a permutation of $\{1, \ldots, r+s\}$, as follows:

Notation. Let $\mathcal{Z}_{r,s} :=$

$$\{\tau \in \mathcal{S}_{r+s} : \tau(1) < \cdots < \tau(r) \quad \text{and} \quad \tau(r+1) < \cdots < \tau(r+s)\}.$$

This has the advantage of letting us avoid a detailed description of what is now $\operatorname{sgn} \tau$, the sign corresponding to a choice $\tau(1) < \cdots < \tau(r)$ of r elements of $\{1, \ldots, r+s\}$. We have the following concise formula.

Note.

$$(\omega \wedge \eta)_{\mu_1 \ldots \mu_{r+s}} = \sum_{\tau \in \mathcal{Z}_{r,s}} \operatorname{sgn} \tau \cdot \omega_{\mu_{\tau(1)} \ldots \mu_{\tau(r)}} \eta_{\mu_{\tau(r+1)} \ldots \mu_{\tau(r+s)}}.$$

Thus the sum has $\binom{r+s}{r}$ summands; if $r = s = 1$, for instance, it has two. The formula for the components of the wedge product of two 1-forms α and β reads

$$(\alpha \wedge \beta)_{\mu\nu} = \alpha_\mu \beta_\nu - \alpha_\nu \beta_\mu.$$

Next we recall the volume form ω_M, for which we computed the following formula in Section 12.3.

Note. *In orientation-preserving local coordinates, the volume form is given by*

$$\omega_M = \sqrt{|g|}\, dx^1 \wedge \ldots \wedge dx^n.$$

Hence its component function is given by $\omega_{1 \ldots n} = \sqrt{|g|}$, *where* $g := \det(g_{\mu\nu})$.

It follows from $\eta \wedge *\zeta = \langle \eta, \zeta \rangle \omega_M$ that, for all $\eta, \zeta \in \Omega^k M$,

$$\sum_{\tau \in \mathcal{Z}_{k,n-k}} \operatorname{sgn} \tau \cdot \eta_{\tau_1 \ldots \tau_k} (*\zeta)_{\tau_{k+1} \ldots \tau_n} = \sum_{\mu_1 < \ldots < \mu_k} \eta_{\mu_1 \ldots \mu_k} \zeta^{\mu_1 \ldots \mu_k} \sqrt{|g|}.$$

Corollary (Star operator in the Ricci calculus). *For* $\zeta \in \Omega^k M$,

$$(*\zeta)_{\tau_{k+1} \ldots \tau_n} = \operatorname{sgn} \tau \cdot \sqrt{|g|} \zeta^{\tau_1 \ldots \tau_k}$$

in orientation-preserving local coordinates.

Of course, this is proved first for $\tau \in \mathcal{Z}_{k,n-k}$, by an appropriate choice of η. But then clearly it also holds for arbitrary $\tau \in \mathcal{S}_n$.

13.9 The Divergence and the Laplacian in the Ricci Calculus

We know from the definition (see the local formula in Section 8.6) how to compute the Cartan derivative in local coordinates. This gives the following formula for the components.

Note (Cartan derivative in the Ricci calculus).

$$(d\omega)_{\mu_1\dots\mu_{k+1}} = \sum_{i=1}^{k+1}(-1)^{i-1}\partial_{\mu_i}\omega_{\mu_1\dots\widehat{\mu_i}\dots\mu_{k+1}}.$$

If we combine this formula and those from Section 13.8 into a general expression for the coderivative in arbitrary coordinates, we get something of a monstrosity, which we hesitate to write down without a special reason. Instead, we take a closer look at the special case $k = 1$.

In this case, the coderivative is defined as

$$\delta = (-1)^{n-1} * d*^{-1} : \Omega^1 M \longrightarrow \Omega^0 M.$$

Since $** = (-1)^{k(n-k)+\operatorname{index} M}\mathrm{Id}_{\Omega^k M}$, as we established in Note 3 in Section 12.3, we also have

$$\delta = (-1)^{\operatorname{index} M} * d * .$$

But for a 1-form $\alpha \in \Omega^1 M$, the formulas above for $*$ and d give

$$(*\alpha)_{1\dots\widehat{\mu}\dots n} = (-1)^{\mu-1}\sqrt{|g|}\,\alpha^\mu,$$

$$(d * \alpha)_{1\dots n} = \sum_{\mu=1}^{n} \partial_\mu(\sqrt{|g|}\,\alpha^\mu).$$

Now it seems like time to apply the $*$-formula (the corollary at the end of the preceding section) again, but this is a bit awkward, and we prefer to observe that we have now computed

$$d * \alpha = \sum_{\mu=1}^{n} \partial_\mu(\sqrt{|g|}\alpha^\mu)dx^1 \wedge \ldots \wedge dx^n$$
$$= \frac{1}{\sqrt{|g|}} \sum_{\mu=1}^{n} \partial_\mu(\sqrt{|g|}\alpha^\mu)\omega_M$$

and already know $*\omega_M = (-1)^{\operatorname{index} M}1$ from Note 2 in Section 12.3. It follows that

$$*d * \alpha = (-1)^{\operatorname{index} M}\frac{1}{\sqrt{|g|}} \sum_{\mu=1}^{n} \partial_\mu(\sqrt{|g|}\alpha^\mu).$$

Using the summation convention, we state this as a corollary.

Corollary. *The coderivative $\delta : \Omega^1 M \to \Omega^0 M$ is described in local coordinates by*

$$\delta\alpha = \frac{1}{\sqrt{|g|}}\partial_\mu(\sqrt{|g|}\alpha^\mu).$$

The function $\delta\alpha$ is also called the ***divergence*** of the vector field $\alpha^\mu\partial_\mu$. For functions (0-forms), the Laplacian $\Delta : \Omega^0 M \to \Omega^0 M$ is defined by $\Delta = \delta d$, or in local coordinates

$$\Delta f = \frac{1}{\sqrt{|g|}}\partial_\mu(\sqrt{|g|}\partial^\mu f).$$

If we expand the Ricci shorthand into an ordinary formula, continuing to write g for $\det(g_{\mu\nu})$ and $(g^{\mu\nu})$ for the inverse of $(g_{\mu\nu})$, we obtain another corollary.

Corollary. *The Laplacian $\Delta := \delta d : \Omega^0 M \to \Omega^0 M$ for functions on a semi-Riemannian manifold M is given in local coordinates by*

$$\Delta f = \frac{1}{\sqrt{|g|}} \sum_{\mu,\nu=1}^{n} \frac{\partial}{\partial x^\mu}(\sqrt{|g|}g^{\mu\nu}\frac{\partial}{\partial x^\nu}f).$$

To illustrate this, we apply the formula to the sphere $M := S^2 \subset \mathbb{R}^3$, with spherical coordinates φ and ϑ. The coordinates are obviously orthogonal: $g_{12} = 0$, hence $g^{21} = 0$ as well. The terms g_{11} and g_{22} are the squares of the velocities of the φ and ϑ coordinate curves, respectively, so $g_{11} = \sin^2\vartheta$ and $g_{22} = 1$. Thus $g = \sin^2\vartheta$, $g^{11} = 1/(\sin^2\vartheta)$, and $g^{22} = 1$.

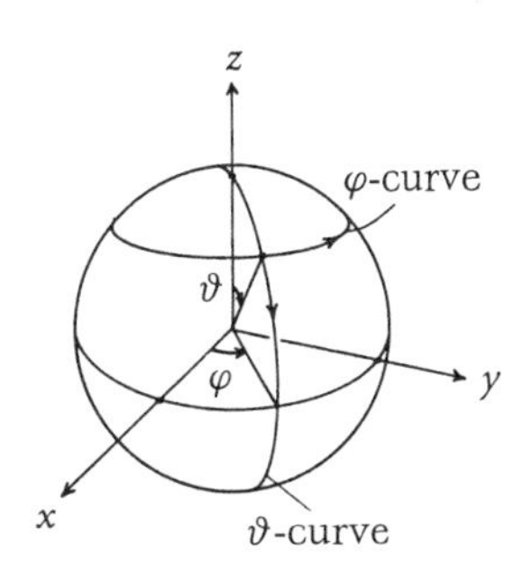

Figure 13.1. Spherical coordinates φ, ϑ on S^2: $x = \sin\vartheta\cos\varphi$, $y = \sin\vartheta\sin\varphi$, and $z = \cos\vartheta$.

Corollary. *In spherical coordinates φ and ϑ, the Laplacian Δ_{S^2} for functions on S^2 is*

$$\Delta_{S^2} = \frac{1}{\sin^2\vartheta}\frac{\partial^2}{\partial\varphi^2} + \frac{1}{\sin\vartheta}\frac{\partial}{\partial\vartheta}(\sin\vartheta\frac{\partial}{\partial\vartheta}).$$

13.10 Concluding Remarks

Every book, or at least every *volume*, must come to an end, and the author takes his leave of the present work by answering a question quite a few readers may have asked themselves already. Why, a reader may ask, does an author who—as he always says himself—values concepts and intuition so highly give so much space to a mere system of notation like the Ricci calculus?

Well, what made me do it is that the conventions of the Ricci calculus are used in the *physics literature*. I would be delighted if an occasional physicist reader were to find my explanations useful. But the explanations are really written for mathematicians. A physics student, I imagine, gets a feel for the calculus by working through concrete problems, and in any case his interest lies in the physical rather than the mathematical content of his formulas. But a mathematician just interested in the abstract geometric aspects of a physical theory, who looks at the physics literature more or less from the outside, as a foreigner, is in a completely different situation.

Whether the physicists' use of the calculus is a mathematical anachronism or the best solution of their notational problems is still, I think, unsettled, but in any case we couldn't read many of the formulas at all without knowing the conventions, and often it's only from the calculus (whose geometric background we know, after all) that we get a hint as to what kind of mathematical objects are really being discussed.

I don't want to reveal here the whole panorama of difficulties awaiting a mathematician who reads physics texts, but I do have to explain something so you won't curse me unfairly when you still can't read every index-studded formula straight off the page.

Namely, you have to be prepared to encounter a great many other kinds of indices besides the actual Ricci indices that refer to space-time coordinates. This comes from the physicists' tendency to choose bases in all vector spaces and, in doing so, introduce indices to which Ricci-like conventions are applied more or less consistently. A major source of such indices is the Lie groups that occur, together with their Lie algebras and the representations of the algebras, in elementary particle physics. Lie groups usually appear at the outset as matrix groups, and their Lie algebras as algebras of matrices (with indices). A basis is chosen in the Lie algebra (this gives one index) and the Lie bracket is described accordingly by structure constants (this gives three indices). A representation assigns matrices to the basis elements, and the matrices have indices referring to the basis of the representation space. Not to mention indices that distinguish among different representations and indices that distinguish among types of particles.

Perhaps in some far-off time this baroque splendor of indices will be discarded, but if we want to listen to the physicists in our time—and they have fascinating things to say—then we have to accept their current language, and a little Ricci calculus is part of that.

13.11 Test

(1) The star operator $*: \Omega^k X \to \Omega^{3-k} X$ for open $X \subset \mathbb{R}^3$ with the usual metric and orientation, viewed in terms of the "translation isomorphisms" as a map $C^\infty(X) \to C^\infty(X)$ for $k = 0, 3$ or as a map $\mathcal{V}(X) \to \mathcal{V}(X)$ for $k = 1, 2$, is

- ☐ the identity on $C^\infty(X)$ or $\mathcal{V}(X)$ for $k = 0, 1, 2, 3$.
- ☐ Id on $C^\infty(X)$ for $k = 0$ and on $\mathcal{V}(X)$ for $k = 2$, but $-$Id on $C^\infty(X)$ for $k = 3$ and on $\mathcal{V}(X)$ for $k = 1$.
- ☐ Id on $C^\infty(X)$ for $k = 0$ and $k = 3$, but $-$Id on $\mathcal{V}(X)$ for $k = 1$ and $k = 2$.

(2) Let M be a manifold, without a metric. Let a linear map $T_pM \to T_pM$ be described in the Ricci calculus by the matrix a_ν^μ (or more precisely by $v^\nu \mapsto a_\nu^\mu v^\nu$) and the dual map $T_p^*M \to T_p^*M$ by b_ν^μ (or $\omega_\mu \mapsto b_\nu^\mu \omega_\mu$ in the sense of the Ricci conventions). Then

☐ $b_\nu^\mu = a_\nu^\mu$. ☐ $b_\nu^\mu = a_\mu^\nu$. ☐ $b_\nu^\mu = (a_\mu^\nu)^{-1}$.

(3) Let M be as above, and let matrices a_ν^μ, b_ν^μ, and c_ν^μ (to be read in the Ricci calculus) describe endomorphisms φ, ψ, and $\psi \circ \varphi$ of either T_pM, the first case, or T_p^*M, the second case. Then

- ☐ $c_\nu^\mu = b_\lambda^\mu a_\nu^\lambda$ in the first case, and $c_\nu^\mu = b_\nu^\lambda a_\lambda^\mu$ in the second.
- ☐ $c_\nu^\mu = b_\nu^\lambda a_\lambda^\mu$ in the first case, and $c_\nu^\mu = b_\lambda^\mu a_\nu^\lambda$ in the second.
- ☐ $c_\nu^\mu = b_\lambda^\mu a_\nu^\lambda$ in both cases.

(4) Does the Kronecker symbol $\delta_{\mu\nu}$ describe a tensor in the Ricci calculus on T_pM?

- ☐ Yes, the identity on T_pM.
- ☐ No. To describe the identity, it would have to be written as δ_μ^ν.

☐ No. $\delta_{\mu\nu}$ doesn't have the right behavior under transformations.

(5) According to the formula

$$(\omega \wedge \eta)_{\mu_1 \dots \mu_{r+s}} = \sum_{\tau \in \mathcal{Z}_{r,s}} \operatorname{sgn}(\tau) \omega_{\mu_{\tau(1)} \dots \mu_{\tau(r)}} \eta_{\mu_{\tau(r+1)} \dots \mu_{\tau(r+s)}}$$

from Section 13.8, the wedge product of a 2-form ω with a 1-form η in the Ricci calculus is $(\omega \wedge \eta)_{\lambda\mu\nu} =$

☐ $\omega_{\lambda\mu}\eta_\nu + \omega_{\mu\nu}\eta_\lambda - \omega_{\lambda\nu}\eta_\mu$.

☐ $\omega_{\lambda\mu}\eta_\nu + \omega_{\nu\lambda}\eta_\mu + \omega_{\mu\nu}\eta_\lambda$.

☐ $\omega_{\lambda\mu}\eta_\nu - \omega_{\lambda\nu}\eta_\mu + \omega_{\mu\nu}\eta_\lambda - \omega_{\mu\lambda}\eta_\nu + \omega_{\nu\lambda}\eta_\mu - \omega_{\nu\mu}\eta_\lambda$.

(6) Now let M be a semi-Riemannian manifold. The isomorphisms

$$\flat : T_pM \xrightarrow{\cong} T_p^*M \quad \text{and} \quad \sharp : T_p^*M \xrightarrow{\cong} T_pM$$

given canonically by the metric are written in the Ricci calculus as

☐ $g^{\mu\nu}$ and $g_{\mu\nu}$, respectively.

☐ $g_{\mu\nu}$ and $g^{\mu\nu}$, respectively.

☐ g_μ^ν for both.

(7) What is g_μ^ν?

☐ $g_\mu^\nu = g_{\mu\lambda} g^{\lambda\nu}$.

☐ $g_\mu^\nu = \delta_\mu^\nu = \begin{cases} 1 & \text{for } \mu = \nu, \\ 0 & \text{otherwise.} \end{cases}$

☐ $g_\mu^\nu = \langle \partial_\mu, \partial_\nu \rangle$.

(8) Let $M = \mathbb{R}^4$, as an oriented Lorentz manifold whose Lorentz metric with respect to the coordinates x^0, x^1, x^2, x^3 is given by

$$(g_{\mu\nu}) = \begin{pmatrix} +1 & & & \\ & -1 & & \\ & & -1 & \\ & & & -1 \end{pmatrix}.$$

Then, by the general formula

$$(*\zeta)_{\tau_{k+1}\dots\tau_n} = \operatorname{sgn}\tau \cdot \sqrt{|g|}\,\zeta^{\tau_1\dots\tau_k},$$

specialized to the action of the star operator on 2-forms $F \in \Omega^2 M$,

- ☐ $(*F)_{01} = F^{23} = F_{23}$. In particular, $*(dx^2 \wedge dx^3) = dx^0 \wedge dx^1$.
- ☐ $(*F)_{01} = F^{23} = F_{23}$. In particular, $*(dx^0 \wedge dx^1) = dx^2 \wedge dx^3$.
- ☐ $(*F)_{23} = F^{01} = -F_{01}$. In particular, $*(dx^0 \wedge dx^1) = -dx^2 \wedge dx^3$.

(9) The divergence $\frac{1}{\sqrt{|g|}}\partial_\mu(\sqrt{|g|}\,v^\mu)$ of a vector field v^μ, in the same coordinates on Minkowski space, is equal to

- ☐ $\partial_0 v^0 + \partial_1 v^1 + \partial_2 v^2 + \partial_3 v^3$.
- ☐ $\partial_0 v^0 - \partial_1 v^1 - \partial_2 v^2 - \partial_3 v^3$.
- ☐ $-\partial_0 v^0 + \partial_1 v^1 + \partial_2 v^2 + \partial_3 v^3$.

(10) Minkowski space again! If we denote the coordinates above by t, x, y, and z, then applying the Laplacian

$$\frac{1}{\sqrt{|g|}}\partial_\mu(\sqrt{|g|}\,\partial^\mu)$$

to a function $f : M \to \mathbb{R}$ gives

- ☐ $\frac{\partial^2}{\partial t^2} f + \frac{\partial^2}{\partial x^2} f + \frac{\partial^2}{\partial y^2} f + \frac{\partial^2}{\partial z^2} f$.
- ☐ $\frac{\partial^2}{\partial t^2} f - \frac{\partial^2}{\partial x^2} f - \frac{\partial^2}{\partial y^2} f - \frac{\partial^2}{\partial z^2} f$.
- ☐ $-\frac{\partial^2}{\partial t^2} f + \frac{\partial^2}{\partial x^2} f + \frac{\partial^2}{\partial y^2} f + \frac{\partial^2}{\partial z^2} f$.

13.12 Exercises

EXERCISE 13.1. Let X be an open subset of $\mathbb{R}^3$ and set $M := \mathbb{R} \times X \subset \mathbb{R}^4$. Intuitively, we picture X as a domain in space and the coordinate t of the factor $\mathbb{R}$ as time. In this exercise and the ones that follow, we want to assimilate the Cartan calculus for the space-time M into our intuition, which separates space and time. Before you can start calculating, though, we have to set things up.

We denote the space of *time-dependent* k-forms on X by $\Omega^k_{\text{time-dep.}} X$ or, a bit more concisely, by $\Omega^k_{\text{t.d.}} X \subset \Omega^k M$. To be precise,

$$\Omega^k_{\text{t.d.}} X := \{\omega \in \Omega^k M : \partial_t \lrcorner\, \omega = 0\}.$$

If we write k-forms on M in the coordinates $x^0 := t$ and x^1, x^2, x^3 of $\mathbb{R}^4$ as

$$\omega = \sum_{\mu_1 < \cdots < \mu_k} \omega_{\mu_1 \ldots \mu_k} dx^{\mu_1} \wedge \ldots \wedge dx^{\mu_k}$$

and sort the summands according to whether $\mu_1 = 0$ or not, we see that every k-form on the space-time M can be represented uniquely as $\omega = dt \wedge \eta + \zeta$, where $\eta \in \Omega^{k-1}_{\text{t.d.}} X$ and $\zeta \in \Omega^k_{\text{t.d.}} X$, and in what follows we will always refer to this isomorphism

$$\begin{array}{ccc} \Omega^{k-1}_{\text{t.d.}} X \oplus \Omega^k_{\text{t.d.}} X & \xrightarrow{\cong} & \Omega^k M, \\ \begin{pmatrix} \eta \\ \zeta \end{pmatrix} & \longmapsto & dt \wedge \eta + \zeta \end{array}$$

in order to bring the space-time forms closer to our intuition.

Three operators act on time-dependent forms on the spatial domain X: the spatial Cartan derivative

$$d_X : \Omega^k_{\text{t.d.}} X \to \Omega^{k+1}_{\text{t.d.}} X,$$

the spatial star operator (with respect to the usual metric on $\mathbb{R}^3$)

$$*_X : \Omega^k_{\text{t.d.}}X \to \Omega^{3-k}_{\text{t.d.}}X,$$

and the partial derivative with respect to time

$$\partial_t : \Omega^k_{\text{t.d.}}X \to \Omega^k_{\text{t.d.}}X.$$

Exercise 13.1 asks you to express the four-dimensional Cartan derivative $d_M : \Omega^k M \to \Omega^{k+1}M$ and the star operator

$$*_M : \Omega^k M \to \Omega^{4-k}M,$$

which refers to the usual orientation and the Lorentz metric on $\mathbb{R}^4$, in terms of d_X, $*_X$, and ∂_t.

EXERCISE 13.2. Now we can go a step further and also interpret the time-dependent forms on X with the usual translation isomorphisms as time-dependent functions or vector fields on X. From the de Rham complex of M, we then obtain a diagram

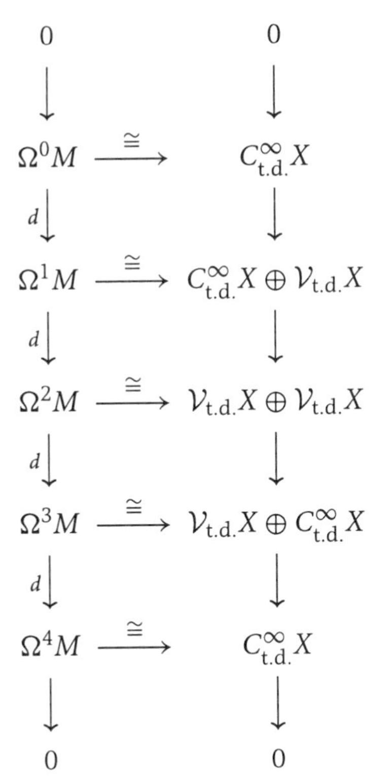

$$\begin{array}{ccc}
0 & & 0 \\
\downarrow & & \downarrow \\
\Omega^0 M & \xrightarrow{\cong} & C^\infty_{\text{t.d.}}X \\
d\downarrow & & \downarrow \\
\Omega^1 M & \xrightarrow{\cong} & C^\infty_{\text{t.d.}}X \oplus \mathcal{V}_{\text{t.d.}}X \\
d\downarrow & & \downarrow \\
\Omega^2 M & \xrightarrow{\cong} & \mathcal{V}_{\text{t.d.}}X \oplus \mathcal{V}_{\text{t.d.}}X \\
d\downarrow & & \downarrow \\
\Omega^3 M & \xrightarrow{\cong} & \mathcal{V}_{\text{t.d.}}X \oplus C^\infty_{\text{t.d.}}X \\
d\downarrow & & \downarrow \\
\Omega^4 M & \xrightarrow{\cong} & C^\infty_{\text{t.d.}}X \\
\downarrow & & \downarrow \\
0 & & 0
\end{array}$$

What happens to the Cartan derivative and the star operator on M when we do this?

EXERCISE 13.3. The formula

$$\int_U \psi \, dS = \iint_G \psi(x, y, z(x, y))\sqrt{1 + (\tfrac{\partial z}{\partial x})^2 + (\tfrac{\partial z}{\partial y})^2}\, dx\, dy$$

for a differentiable function $z = z(x, y)$ of *two* variables was given in Section 10.8. Prove the obvious generalization to the case of a function $f = f(x^1, \ldots, x^n)$ of n variables.

EXERCISE 13.4. Prove the formula

$$d(X \lrcorner\, \omega_M) = (\operatorname{div} X)\omega_M$$

for the divergence (defined in Section 13.9) of a vector field on an oriented semi-Riemannian manifold.

13.13 Hints for the Exercises

FOR EXERCISE 13.1. You should find that the following diagrams are commutative:

$$\begin{array}{ccc}
\Omega^k M & \xrightarrow{\quad d \quad} & \Omega^{k+1} M \\
\cong\uparrow & & \cong\uparrow \\
\Omega^{k-1}_{\text{t.d.}} X \oplus \Omega^k_{\text{t.d}} X & \xrightarrow[\begin{pmatrix} -d_X & \partial_t \\ & d_X \end{pmatrix}]{} & \Omega^k_{\text{t.d.}} X \oplus \Omega^{k+1}_{\text{t.d.}} X
\end{array}$$

and

$$\begin{array}{ccc}
\Omega^k M & \xrightarrow{\quad *_M \quad} & \Omega^{4-k} M \\
\cong\uparrow & & \cong\uparrow \\
\Omega^{k-1}_{\text{t.d.}} X \oplus \Omega^k_{\text{t.d.}} X & \xrightarrow[\begin{pmatrix} & *_X \\ (-1)^{k-1} *_X & \end{pmatrix}]{} & \Omega^{3-k}_{\text{t.d.}} X \oplus \Omega^{3-k+1}_{\text{t.d.}} X
\end{array}$$

FOR EXERCISE 13.2. In the classical electrodynamics of the vacuum, the units can be chosen so that only three time-dependent vector fields and one time-dependent function on $X \subset \mathbb{R}^3$ need be considered, namely

the electric field strength $\vec{E}$,
the magnetic induction $\vec{B}$,
the current density $\vec{J}$, and
the charge density ρ,

and so that Maxwell's equations read

$$\begin{aligned}
&\operatorname{curl} \vec{E} = -\dot{\vec{B}},\\
&\operatorname{div} \vec{B} = 0,\\
&\operatorname{curl} \vec{B} = -\dot{\vec{E}} + \vec{J},\\
&\operatorname{div} \vec{E} = \rho.
\end{aligned}$$

If $\begin{pmatrix} -\vec{E} \\ \vec{B} \end{pmatrix}$ is translated into a 2-form $F \in \Omega^2(\mathbb{R}\times X)$, the ***Faraday tensor***, and $\begin{pmatrix} -\vec{J} \\ \rho \end{pmatrix}$ into a 3-form $j \in \Omega^3(\mathbb{R}\times X)$, the ***four-current density***, then Maxwell's equations become

$$\begin{aligned}
dF &= 0,\\
d * F &= j,
\end{aligned}$$

and the equation $dj = 0$ that follows from $d * F = j$ becomes the ***continuity equation*** $\operatorname{div} \vec{J} + \dot{\rho} = 0$.

It isn't by chance that Maxwell's equations become so simple in the Cartan calculus of Minkowski space $\mathbb{R}^4$, but to go into more detail I would have to give more background than opportunity permits.

FOR EXERCISE 13.3. According to the formula for the volume form in Section 13.8, solving this problem is mainly a question of finding the determinant of the symmetric matrix with components

$$g_{\mu\nu} = \delta_{\mu\nu} + \partial_\mu f \cdot \partial_\nu f.$$

But this is the product of the eigenvalues, counting multiplicities. As a self-adjoint operator on $\mathbb{R}^n$, the matrix is easy to understand: it's the sum of the identity and an operator of rank one, and the eigenvalues are visible to the naked eye.

FOR EXERCISE 13.4. The coordinate formula in Section 13.9 already showed that the only influence of the metric on taking the divergence of a vector field comes in through the volume form. The assertion of Exercise 13.4 offers a coordinate-free interpretation of this situation.

14 CHAPTER

Answers to the Test Questions

Problem 1

1	2	3	4	5	6	8	9	10	11	12	13
		×			×			×			×
	×		×		×				×	×	
×		×		×		×	×			×	

Problem 2

1	2	3	4	5	6	8	9	10	11	12	13
×	×	×					×	×		×	×
		×	×	×							
					×	×			×		

Problem 3

1	2	3	4	5	6	8	9	10	11	12	13
×	×				×	×					×
				×			×	×	×	×	
		×	×	×							

	1	2	3	4	5	6	8	9	10	11	12	13
Problem 4		×	×		×					×	×	
						×			×			×
	×			×			×	×				×

	1	2	3	4	5	6	8	9	10	11	12	13
Problem 5	×					×				×		×
		×		×			×	×			×	
			×		×	×			×			

	1	2	3	4	5	6	8	9	10	11	12	13
Problem 6		×	×			×					×	
				×	×			×	×			×
	×						×			×		

	1	2	3	4	5	6	8	9	10	11	12	13
Problem 7					×	×	×				×	×
	×		×				×			×		×
		×	×	×		×		×	×	×	×	

	1	2	3	4	5	6	8	9	10	11	12	13
Problem 8		×					×			×		×
	×			×	×			×	×	×	×	
	×		×		×	×				×	×	×

	1	2	3	4	5	6	8	9	10	11	12	13
Problem 9	×		×			×	×		×		×	×
		×			×							
				×				×		×		

Problem 10

1	2	3	4	5	6	8	9	10	11	12	13
			×				×		×	×	
	×	×		×							×
×		×			×	×		×			

Bibliography

[AM] R. Abraham and J. E. Marsden. *Foundations of Mechanics.* 2nd ed., Addison-Wesley, Reading, Mass., 1978.

[BJ] T. Bröcker and K. Jänich. *Introduction to Differential Topology.* Cambridge University Press, Cambridge, 1982.

[C] H. Cartan. Les travaux de Georges de Rham sur les variétés différentiables. In A. Haefliger and R. Narasimhan, editors, *Essays on Topology and Related Topics. Mémoires dédiés à Georges de Rham*, Springer-Verlag, Berlin–Heidelberg–New York, 1970.

[J:*Lin*] K. Jänich. *Linear Algebra.* Springer-Verlag, Berlin–Heidelberg–New York, 1994.

[J:*Top*] K. Jänich. *Topology.* Springer-Verlag, Berlin–Heidelberg–New York, 1984.

[W] F. W. Warner. *Foundations of Differentiable Manifolds and Lie Groups.* Scott, Foresman and Company, Glenview, Illinois-London, 1971.

Index

Undergraduate Texts in Mathematics

(continued from page ii)

Hartshorne: Geometry: Euclid and Beyond.
Hijab: Introduction to Calculus and Classical Analysis.
Hilton/Holton/Pedersen: Mathematical Reflections: In a Room with Many Mirrors.
Iooss/Joseph: Elementary Stability and Bifurcation Theory. Second edition.
Isaac: The Pleasures of Probability. *Readings in Mathematics.*
James: Topological and Uniform Spaces.
Jänich: Linear Algebra.
Jänich: Topology.
Jänich: Vector Analysis.
Kemeny/Snell: Finite Markov Chains.
Kinsey: Topology of Surfaces.
Klambauer: Aspects of Calculus.
Lang: A First Course in Calculus. Fifth edition.
Lang: Calculus of Several Variables. Third edition.
Lang: Introduction to Linear Algebra. Second edition.
Lang: Linear Algebra. Third edition.
Lang: Undergraduate Algebra. Second edition.
Lang: Undergraduate Analysis.
Lax/Burstein/Lax: Calculus with Applications and Computing. Volume 1.
LeCuyer: College Mathematics with APL.
Lidl/Pilz: Applied Abstract Algebra. Second edition.
Logan: Applied Partial Differential Equations.
Macki-Strauss: Introduction to Optimal Control Theory.
Malitz: Introduction to Mathematical Logic.
Marsden/Weinstein: Calculus I, II, III. Second edition.
Martin: The Foundations of Geometry and the Non-Euclidean Plane.
Martin: Geometric Constructions.
Martin: Transformation Geometry: An Introduction to Symmetry.
Millman/Parker: Geometry: A Metric Approach with Models. Second edition.
Moschovakis: Notes on Set Theory.
Owen: A First Course in the Mathematical Foundations of Thermodynamics.
Palka: An Introduction to Complex Function Theory.
Pedrick: A First Course in Analysis.
Peressini/Sullivan/Uhl: The Mathematics of Nonlinear Programming.
Prenowitz/Jantosciak: Join Geometries.
Priestley: Calculus: A Liberal Art. Second edition.
Protter/Morrey: A First Course in Real Analysis. Second edition.
Protter/Morrey: Intermediate Calculus. Second edition.
Roman: An Introduction to Coding and Information Theory.
Ross: Elementary Analysis: The Theory of Calculus.
Samuel: Projective Geometry. *Readings in Mathematics.*
Scharlau/Opolka: From Fermat to Minkowski.
Schiff: The Laplace Transform: Theory and Applications.
Sethuraman: Rings, Fields, and Vector Spaces: An Approach to Geometric Constructability.
Sigler: Algebra.
Silverman/Tate: Rational Points on Elliptic Curves.
Simmonds: A Brief on Tensor Analysis. Second edition.
Singer: Geometry: Plane and Fancy.
Singer/Thorpe: Lecture Notes on Elementary Topology and Geometry.
Smith: Linear Algebra. Third edition.
Smith: Primer of Modern Analysis. Second edition.
Stanton/White: Constructive Combinatorics.

Undergraduate Texts in Mathematics

Stillwell: Elements of Algebra: Geometry, Numbers, Equations.
Stillwell: Mathematics and Its History.
Stillwell: Numbers and Geometry. *Readings in Mathematics.*
Strayer: Linear Programming and Its Applications.
Thorpe: Elementary Topics in Differential Geometry.
Toth: Glimpses of Algebra and Geometry. *Readings in Mathematics.*
Troutman: Variational Calculus and Optimal Control. Second edition.
Valenza: Linear Algebra: An Introduction to Abstract Mathematics.
Whyburn/Duda: Dynamic Topology.
Wilson: Much Ado About Calculus.